# SEX, DRUGS

### AND

# ROCK 'N' ROLL

**E CORMIER** is a freelance journalist, science writer and public
eaker with a background in biology. Zoe is also the Head of
mmunications for UK science outreach organisation Guerilla
ence. Originally from Toronto, she now lives in London.

**GUERILLA SCIENCE** create events and installa-
tions for festivals, museums, galleries, and other
cultural partners. They are committed to connecting people with science in new ways, and
producing live experiences that entertain, inspire,
challenge and amaze.

# SEX, DRUGS

## AND

# ROCK 'N' ROLL

The science of hedonism
And the hedonism of science

## Zoe Cormier

**P**

PROFILE BOOKS

First published in Great Britain in 2014 by
PROFILE BOOKS LTD
3A Exmouth House
Pine Street
London EC1R OJH
*www.profilebooks.com*

10 9 8 7 6 5 4 3 2 1

Printed and bound in Great Britain by Clays, Bungay, Suffolk

A CIP catalogue record for this book is available from the British Library.

ISBN 978 1 78125 092 1
eISBN 978 1 84765 949 1

The paper this book is printed on is certified by the © 1996 Forest Stewardship Council A.C. (FSC). It is ancient-forest friendly. The printer holds FSC chain of custody SGS-COC-2061

FSC
www.fsc.org
MIX
Paper from
responsible sources
FSC® C018072

For my parents.

Because if it were not for their mutual appreciation for all three of these things, this book would not exist – and neither would I.

# CONTENTS

# DRUGS

# ROCK 'N' ROLL

*Sex and drugs and rock and roll*
*Are all my mind and body need*
*Sex and drugs and rock and roll*
*Are very good indeed.*

'Sex and Drugs and Rock and Roll'
Ian Dury and the Blockheads

# ACKNOWLEDGEMENTS

First and foremost to Jen Wong, the captain of our ship, an eternal source of strength and inspiration. One of a kind.

For Mark Rosin, co-founder of Guerilla Science. With a PhD in astrophysical mathematics from Cambridge, you had the singular insight that we could turn the art of making science hedonistic into a viable business. You are a genius.

For Jenny Jopson, Louis Buckley and Olivia Koski. All bona fide and brilliant scientists who chose to transform your empirical skills into incandescent careers as entertainers and producers. You are all science rock stars.

For Richard Bowdler, prodigal son of Guerilla Science, who helped found our company in the first place, you uttered the most profound, succinct and inspiring quip on the human condition I have ever heard: 'We exist.'

For all my friends who have drug tendencies and dependencies – both the dead and the living: I love you. Please stop. I miss who you used to be. These chemicals are supposed to be treats. Not lifestyles.

For all my friends and family who have overcome their dependencies. I am proud of you, I love you, and I am grateful that you are still here with me.

For my grandfather, actor and playwright Don Harron. When

I told you I wanted to study science and not literature, you took it in your stride, but said, firmly, 'Fine. But don't ever forget: You are not a scientist – you are a writer.'

For my father, rock promoter Gary Cormier, who taught me more about music than any instructor at the Royal Conservatory ever did.

For my mother, Martha Harron, who is still the best wordsmith I have ever had the good grace to meet.

For my sister Amy, and your strength, prowess, and invaluable advice – in all corners of life. Without you this could not have happened.

For my brother Ben, computational genius and the most generous tech support imaginable. You might detest magnetic tape, but I still treasure the cassette of techno you gave me in 1993. I wouldn't be who I am without you.

For my aunts Kelley and Mary. Everyone needs role models. You are mine.

For my musical pharmacists Duncan Thornley, Kier Wiater-Carnihan, Dan Hartrell, Florian Tanant and Daniel Garber. You enrich my aural landscape as well as my life. Worshipful gratitude to Gemma Wain, Daniel Farrell, Neon Kelly and Rebecca Burn-Callander – writers and unparalleled punners – for feeding my insatiable love of words and making the point of writing worth remembering. Peter Farrel and Debora Malvezzi: you make life worth living.

And ... For England. Your raucous music festivals, riotous drug use, and insatiable lust for life provided the inspiration for this book. Plus, your miserable grey skies made it easy for me to stay indoors and write. You can self-denigrate as much as you like, but to me, you are irrefutably beautiful. Thank you.

# FOREWORD

Six years ago a group of scientists decided to revolutionise how people experience science.

Why? Because science had been left languishing in the doldrums, closeted behind the barriers of privilege, stuffy stereotypes, and obscure jargon meant to intimidate rather than inform. We wanted to set science free and inspire the curiosity and wonder at the world that had so captivated us. We sought to challenge preconceptions of science, and scientists, as unintelligible, dry, dull men in white coats. This was the birth of Guerilla Science: a science collective practising science by stealth, bringing sweet facts to surprised audiences and breathing mind-blowing experiences into places where science was unexpected.

Our playful revolution was born in a bewildering range of venues which became our second homes: abandoned warehouses, Victorian workhouses-cum-psychiatric hospitals, the empty shell of Battersea Power Station, the glitzy caverns of Selfridges, and the inimitable muddy music festival field. We labelled people with stickers proclaiming 'I am stardust', took them on a safari of the fundamental particles of the universe, led them on an auditory tour of the universe inside our heads, and then on a sonic adventure through our galaxy and beyond. We conducted neurological experiments on agreeable punters on a floating island constructed from hay. We made and dissected jelly brains, feeding them to intellectually hungry diners. We even simulated the experience of being a human lab rat subjected to a battery of sensory tests within a giant maze.

Our mission is to explore science in unique, creative ways and produce experiences that inform, entertain and amaze. Because of what we do, people experience science in unorthodox ways, helping them to discover the unimaginable realities that science reveals to us.

As playfulness is evident in all that we do, the science of pleasure and its importance to our everyday lives makes a fitting book. This investigation of the science of hedonism (and the hedonism of science) presents science as a tool for understanding, questioning and, ultimately, appreciating what it truly means to exist in the first place. It will take you on a journey showing how science has been used to understand ourselves over the ages.

Enjoy the ride.

Jen Wong
*Guerilla Science Creative Director*

# INTRODUCTION

What makes us special? *Homo sapiens* may not be the pinnacle of evolution – but we are certainly singular. There is no living thing quite like us. But what exactly is it that makes our species so unique?

The question of just what makes us who we are has occupied our complex, anxious brains since the moment we became self-aware. It is a scientific puzzle, and furiously difficult to answer.

What does it mean to be human?

Is it consciousness? Self-awareness? Mathematical prowess? Linguistic adeptness?

Our name tells us a fair amount about what we think of ourselves: *Homo sapiens*: Latin for 'wise man'. The thinky monkey. We think we're so clever.

Since we put pen to paper, we have aggrandised our 'higher' cognitive capacities as the keys to what demarcates us from the rest of the animal kingdom. We can write things down, add numbers together, think about the past, present and future, and worry ourselves to death over everything we have the capacity to consider.

Moreover, our unrivalled mind may be what makes us human, but it is also what makes us profoundly miserable, anxious and confused. Perhaps we connect with something deeper within ourselves by losing it. Do sex, drugs and rock 'n' roll constitute the three best ways to lose your mind? And does thinking less do us some good now and then?

Meanwhile, our supposedly 'base' pursuits – getting laid, getting high, rocking out – have been relegated to the bin of 'animalistic' impulses. Sex, drugs, and music have been denigrated as primal instincts: pleasurable and powerful, but biologically unimpressive. Animals also copulate, self-dose and make noise. Surely the written word and the mathematical graph are more impressive evolutionary achievements?

But not all human cultures possess writing, the number zero, agriculture, architecture, or many of our supposedly more impressive achievements. Yet all human cultures are united by a uniform embrace of 'base' pursuits. We all enjoy red-hot ruts, take filthy drugs, and make riotous sounds. Sex, drugs and rock 'n' roll are universal characteristics of the human condition. We might like to deny that we are drawn to them, but we are undeniably hooked. And have been since the beginning.

We deem such impulses 'hedonistic'. There are many definitions of the word hedonism, but the phrase 'sensual self-indulgence' fits quite nicely. Saucy treats which are irresistibly alluring yet not necessary for survival.

But if they are universal surely our hedonistic tendencies must teach us something about what it means to be human?

Over the past four centuries scientists have revealed that our sexual, narcotic and musical characteristics are biologically unsurpassed.

We are one of the few mammals with a fleshy penis lacking a bony support, which means that the mind – complex beast it is – is integral to our capacity to copulate. On the other side of the gender coin, the clitoris seems to be the only organ designed for pleasure alone. It contains more nerve endings than any other corner in the body. It is at the same time ancillary, inessential and spectacular. Its presence in our species is noteworthy.

Perhaps most striking: orgasm. It does not seem that anything that has ever graced God's green earth has ever experienced the height and duration of orgasmic experience as we humans.

We are lucky.

Politicians and self-righteous tee-total pontificators might intone us to stay away from drugs, but our evolutionary heritage has instructed us to do otherwise. The hallucinogens in magic mushrooms and LSD lock into our body's keyholes for serotonin more easily than serotonin itself. We might never have discovered our body's own soothing painkillers, the endocannabinoids, were it not for our everlasting love for weed. And the morphine produced from the poppy plant remains the most powerful anaesthetic ever devised – no laboratory technician has ever beaten it in the contest to produce the world's best painkiller. As usual, Plants 1: Humans 0. We do drugs thanks to millions of years of botanical burglary.

The predisposition to take drugs, no matter how lethal or poisonous, is part of what makes us who we are. It is integral to the human condition. Sex and drugs can both trigger the release of exhilarating biochemicals. And so can music. Your favourite tunes could send shivers up your spine even more powerfully than a new lover or a crusty chemical.

Music – the strange, amorphous thing that it is – holds even more intriguing and mysterious clues to the roots of our humanity. Not only could the oldest human creation ever recovered be an instrument – a 40,000 year old flute – ochre paintings indicate we scoured caves as sound technicians in our earliest prehistoric days. Music is ancient, and spectacularly specific to our species. We would not be human were it not for our insatiable desire to crank up the volume.

Sex, drugs and music have all served as divining rods that led us to understand what makes our species the way it is. The pages of this book are filled with countless and exuberant examples of what scientists have discovered: why sex, drugs and music are important. Truth is stranger than fiction, and biology is art.

That sex, drugs and music are crucial components of human nature is probably intuitive to most. But here's the less obvious truth: were it not for our supposedly 'base' impulses, we never would have achieved many ground-breaking scientific discoveries. Hedonism has been integral to intellectual progress.

This is a story about scientists and their craft: a story of rebellion. There is redemption in disobedience.

Reprobation is redeeming. We might never have tracked down many of the most important chemical messengers of the mind were it not for our relationship with illegal narcotics and what they taught us about how the brain works. We understood nicotine before acetylcholine, caffeine before adenosine riphosphate, opium before endorphins, and hallucinogens before serotonin.

Moreover, mind-bending drugs led to profound intellectual insights. Were it not for LSD, we might not know how to unravel the language of DNA with the ease we can today. Psychedelics may have also been integral to the quantum physics renaissance. Nobel Prize winners have attributed their discoveries to mind-bending psychedelics.

This is not just a story about what scientists have discovered, or why naughty treats made us who we are. This is a story about scientists and their craft, and how hedonistic impulses inform our highest pursuits. How the renegades of science have illuminated the secrets behind our deepest impulses. To deny our hedonistic instincts is intellectually stunting.

Science is ubiquitously stereotyped as an uninspired reduction of life's exhilarating complexities into humdrum mundanities.

Take something special, mark it on a graph, and debase it into an insipid uninspired scatterplot of dots and lines, robbing the human condition of all subtlety and nuance.

Nothing could be further from the truth. Not only has science taught us how important our hedonistic impulses are, but in many ways, science would have never progressed were it not for these insatiable drives in the first place.

# SEX

# EVOLUTION'S CLIMAX

For the vast majority of animals, the act of reproduction is fairly banal: she spreads her eggs over the ground, he spreads his sperm over them, and they both walk away without further contact or responsibility. Most fish, frogs, insects and, of course, plants create more of themselves without anything beyond a cursory spurt.

The main purpose is simple, and one of the mechanisms behind all evolutionary change. When making more little members of your species, rather than just creating an exact copy of yourself (cloning – still habitually practised by many micro-organisms, insects and plants), you might do better by shuffling the genetic deck of cards. By mixing your genes with some-body else's, you potentially leave behind descendants that are an improvement on yourself: bigger, stronger, faster, smarter. In other words: better adapted to their environment. That's the ulti-mate reason for procreation in the first place: taking a chance by throwing DNA's dice.

In most species, reproductive roulette is done without pomp and circumstance. Spread, spaff, done. But things just get weirder, and more interesting, from there. Evolution is the mother of invention, and sexual reproduction has been the engine driving the formation of some of life's most bizarre behaviours, chem-istries and anatomies. The shape of sex in the animal kingdom ranges from the ridiculous to the sublime.

And sometimes the very awful.

Rape is the habitual mode of reproduction for a few species of duck.[1] Only 3 per cent of bird species boast external male geni-tals. Of the few that do, rape (or 'forced copulation') frequently results in death of the female by drowning. Males of the common

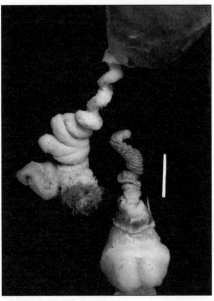

1. Common mallard (*Anas platyrhynchos*) females
contain vaginas that swirl, and males in turn bear penises
that twirl. Male on the right, female on the left.

mallard *Anas platyrhynchos** (Latin for 'broad-snouted duck')
bear contorted, spiralled penises. Female ducks in turn contain
swirling vaginal structures designed to thwart male attempts to
sire offspring. American biologist Dr Patricia Brennan estimates
only 4 per cent of forced copulations result in ducklings (60 per
cent of matings in this species of duck are consensual).[2]

---

\* A quick explanation of the name *A. platyrhynchos*: All species, such as
our own, *Homo sapiens*, are classified by biologists in a Latinised two-word
system known as 'binominal nomenclature'. Some titles are indisputably
pretty, such as *Pipistrellus pipistrellus*, the common brown bat of England. Its
relative, wearing the less pretty name *Myotis lucifugus*, the little brown bat
of North America, features a strange two-stage mating system. In the 'active'
system both partners are awake. In the 'passive' system males copulate with
sleeping bats – regardless of their sex.

Female mallard vaginas turn and twist, peppered with numerous cul-de-sacs which they use to divert an assailant's sperm into dead ends. Researchers speculate that the two genital structures have co-evolved in an escalating 'arms race'.

Female praying mantids famously devour the heads of their lovers – who continue to copulate unabated. 'Love darts' are a common weapon of romantic warfare in slugs and snails. Males pierce holes in the side of a female's body before shoving his sperm haphazardly into her abdominal cavity. Hyena females are burdened with a birth canal that extends directly through the clitoris, and are thus required not only to fornicate but also to give birth through the organ, which is so large it has been dubbed a 'pseudopenis' by zoologists.

Even familiar species mate in bizarre fashions. The penises of cats bear barbs about a millimetre long formed of keratin (the same material that constitutes your fingernails). Adding insult to injury, the barbs point backwards. Interestingly, these do not deter females from seeking male attention. Lionesses, for example, may mate up to a hundred times a day during the fertile portion of their cycle. Cat tongues are also covered in sharp keratinous shards, which one would think would make the act of oral sex rare. But it does occur elsewhere in the animal kingdom. Hyenas have been known to indulge in oral sex, along with gibbons and goats.[3]

Feline penile spines are designed to displace the sperm of other males who have sojourned with the same female. Dislodging the competition is a common function of elaborate phallic structures throughout the animal kingdom (including possibly ourselves). Curiously, the scraping of spines inside the feline vaginal canal is *required* to induce the hormonal transformations that spur ovulation and allow fertilisation to occur. Hence the term for cats and other animals that do not have cyclic ovulation (like us) but need some form of physical or chemical trigger: 'induced ovulators'. Kittens are the product of vaginal lacerations.

But the very worst kind of animal sex I can think of is found in the deep-sea squid *Moroteuthis ingens*. Males dart females with sperm-filled bullets called spermatophores. This is nothing special: most species of squid transfer sperm from male to female through specialised capsules, usually inserted directly into her receptive cavity. But what makes the mode of reproduction in *M. ingens* particularly alarming is that they don't shove their packets into a specialised genital opening on the female's body. Or even on the side of her body near her eggs. They stick the capsules anywhere and everywhere.

These scattered spermatophores then burrow into the female's body. The story of how researchers unravelled the mechanics of this love act is rather memorable. Dutch scientists aboard the research vessel *Dorada* of the Falkland Islands Government Fisheries Department discovered the secrets of deep squid sex[4] by sticking a live male's sperm capsules on to the body of a dead male. It is noteworthy that no female squid took part in this act of scientific sexual investigation. Here is the unsettling account straight from their 2007 paper from the scientific journal *The Biological Bulletin*:

> The everting ejaculatory apparatus has the first contact with the tissue and may facilitate adhesion or the first penetration into the tissue, perhaps by mechanical means. After eversion, the cement body is exposed and may dissolve (perhaps with the aid of proteolytic enzymes) the host tissue to allow further penetration of the spermatangium.

Translation: the sperm capsules appear to be coated in

protein-dissolving enzymes specifically engineered to melt through the flesh of the female, allowing sperm to enter her body cavity.

The *coup de grâce*: once they have burrowed into her tissue, *M. ingens* sperm are then able to travel from anywhere in the female straight to her eggs, burrowing this way and that. So males understandably aren't too fussed about where they dart the female with their load.

'When we started pulling female squid out of the ocean covered in these strange white things, we thought they were parasites,' recalls squid researcher Dr Inger Winkelman,[5] who has studied similar behaviour in the legendary giant squid *Architeuthis*, which also spear females with sperm in a similar but slightly more accurate fashion.

This spectrum of reproduction in animals teaches us that human sex is a special thing. We derive enormous – probably biologically unrivalled – pleasure from the act. We have learned how to do it without the responsibility of procreation. And we can go for great lengths of time. Even our randy cousins the bonobos, *Pan paniscus*, only copulate for a fraction of the time we do. An average bout of bonobo thrusting only lasts fifteen seconds, and our cousins the chimps can only sustain a mere seven.[6]

Whether or not the human species is the 'pinnacle' of evolution is a matter of opinion. But human sex is truly one of evolution's greatest achievements. The happiness, contentment and joy that can be gleaned from the act is pretty hard to capture in words, though many have tried. Hemingway's description of 'the earth moving' is pretty good.

It's unlike anything else we experience. It's free and, if the right partners are available, just about limitless. There may be some truth to the idea that 'life is suffering', but frankly, sex is one of the things that makes life worth living. The experience of being alive is fraught with many unpleasant ordeals, from sickness to depression and eventually death, but sex is one of the human condition's most redeeming qualities. The all-consuming ecstasy of the human orgasm took 3 billion years of evolution to appear. It is a gift.

So why, then, is the history of human sexuality such a miserable, sordid, horrible affair? Castrations, clitorectomies, incarcerations, shame, celibacy, mutilations, expulsions and executions. Since we came down from the trees we have found excuses to lock each other up, call one another decrepit and diseased, and even mutilate and kill one another over how we decide to use our bodies to feel pleasure. Secularists like to point fingers at religious authorities for instilling shame and encouraging persecution, but scientists have also been guilty of justifying prejudice and cruelty. Just because an idea derives from the supposedly unbiased gathering of data rather than from divine inspiration or impassioned conviction, does not guarantee it will be humane or tolerant. Homosexuals were categorised as mentally disturbed by the world's smartest psychiatrists until 1973. Skilled western surgeons have designed tools to mutilate the clitoris. Women who do not climax from penetration alone have been repeatedly deemed immature or diseased.

A book exploring the scientific study of sex, drugs and rock 'n' roll ought to be fun. And no subject should be more fun than sex: it is one of the best things about being human. Moreover, unlike drugs, too much of it probably won't hurt you (barring a sexually transmitted infection). In fact, studies indicate that regular sex is very good for you. But a story that should be almost invariably a happy one is instead riddled with persecution, misery, self-denial, self-loathing, disfigurement and hatred.

Why do we do this to each other – and to ourselves? Why is the supposedly rational human animal so prone to irrational and poisonous self-denigration? Moreover, what does this ultimately reveal about how our primal instincts influence intellectual pursuits? The body's subconscious impulses clearly hold more sway over our supposedly cerebral conclusions than we might ever wish to concede.

# THE HUMAN ANIMAL

We are constantly reminded that the pile of electrified jelly sitting inside our skulls is 'the most complex object in the universe', and neuroscientists obsess over what makes our humming human brains so special. Is it language? Consciousness? Is it – as we will later explore – music?

As special as our brains may be, we often fail to see how remarkable our bodies are: our behaviours, anatomies and experiences. Many aspects of the human form are – let's be honest – a bit dull compared to other living things. We lack the ability to fly, glow, or live for 10,000 years.* But we do boast some spectacular sexual idiosyncrasies. These tell us that our sexuality could be an important piece in the puzzle of what makes us human. Why we are the way we are.

Let's start with the penis. While the size of a man's member continues to be a great source of anxiety for many, the size of the human penis is in fact remarkably large compared to other primates. Silverback male gorillas – easily capable of breaking the neck of even the strongest man alive – only wield penises four centimetres in length. In fact, humans have the largest penis of any primate – both in absolute terms (total length) and relative terms (in proportion to body size). As influential sex scientist

---

* Or, 9,550 years to be exact, the estimated age of Old Tjikko, a Norway spruce (*Picea abies*) still ticking in Sweden.

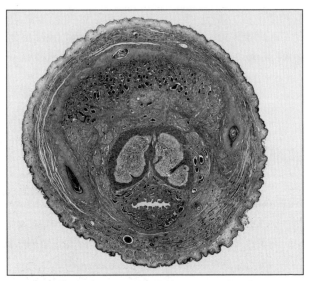

2. The human penis contains porous chambers,
the corpus cavernosa, which are designed to swell
with blood and produce an erection.

Geoffrey Miller put it: 'Adult male humans have the longest, thickest, and most flexible penises of any living primate.'[7]

Intriguingly, our penises are unusually soft. Almost all mammals possess a penis bone, including cats, dogs, whales, and all other primates including chimpanzees. But ours are flaccid and squishy (until required). A 'boner' is in fact boneless, and we are unusual in this respect.

Why should this be so? Penis bones allow for quickie mating: males are capable of penetrating females at any time. The lack of calcified material in our own species means that other triggers are required to produce and sustain an erection.

We have Leonardo da Vinci (1452–1519) to thank for elevating our understanding of the erection: he was the first to correctly identify blood as the physiological mechanism responsible for inflating the penis. Prior to his insight, it was assumed that

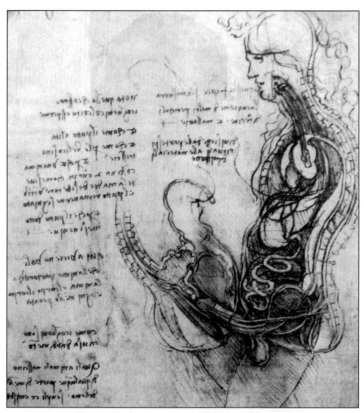

3. Leonardo Da Vinci, *The Copulation* (1493): 'I expose to men the origin of their first, and perhaps second, reason for existing.'

the organ swelled with air. Meticulous dissection of the phallus reveals three spongy cavities, two that run in parallel, the corpus cavernosa, and one narrow chamber underneath them, the corpus spongiosum, all of which flood with blood and permit penetration.

This spongy nature of the human penis allows for another anatomical quirk, which Leonardo was not capable of elucidating, but which modern scientific instruments have. Through his work with cadavers in 1493 he produced one of the most famous illustrations of human intercourse, dubbed simply: 'The Copulation'.

'I expose to men the origin of their first, and perhaps second, reason for existing,' the master wrote. Note that channels connect the breasts to the uterus, and the skull to the penis, reflecting the ancient beliefs that breast milk derives from menstrual blood and semen stems from the brain. As Greek philosopher Pythagoras (c. 570–c. 495 BC) colourfully put it 2,000 years earlier, ejaculate is a 'clot of brain-containing hot vapour'. The disproportionate representation of the female body compared to the male in da Vinci's illustration is worth noting.

This artistic and scientific work is made all the more interesting by virtue of the fact that Leonardo described the act of sex as 'disgusting'. Though we can be certain of little about his sex life, it is presumed by many that he never engaged in the act with a woman. Nonetheless, his sketch reflects what most of us would assume: the penis slides straight into the receptive canal of a woman like a sword into a sheath.

This belief held sway for centuries. Pioneering obstetrician and gynaecologist Robert Latou Dickinson (1861–1950) also concluded that the penis drives straight into the vagina based on his own examinations with glass dildos, and illustrated the pairing of the sexes in the same fashion in 1933.[8]

It was more than 500 years after Leonardo pencilled his influential sketch that scientists could examine the act of copulation in cross section as it was actually happening. To their surprise, they discovered that the penis does not impale: it bends. Dutch researcher Pek Van Andel, inspired by an analysis he performed in 1991 on the vocal chords of a singer, wondered if modern scientific imaging technologies could uncover anything useful or informative about the pairing of human genitals. A 1992 study[9] using ultrasound – probing the innards with high-frequency

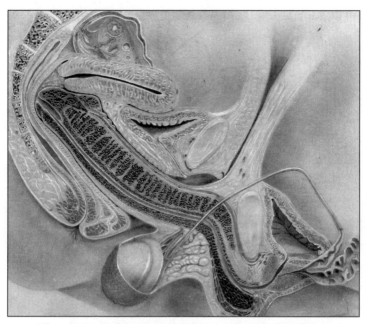

4. Obstetrician Robert Latou Dickinson illustrated in this very convincing artwork what he believed the penis would look like nestled inside a woman: straight and unwavering.

sound waves (akin to the sonar of bats and the radar of ships) – produced fuzzy images that illuminated little.

Van Andel hoped magnetic resonance imaging (MRI) could improve upon the ultrasound images: MRI scanners (often referred to as 'doughnuts') are enormous circular chambers and the workhorse of medical imaging. Gigantic magnets move around the patient, who lies within the claustrophobic scanner. The magnetic fields excite the hydrogen atoms in water molecules (the H in $H_2O$), and sophisticated cameras document the atoms' frenzied jostling.

Why ask a couple to copulate inside a wheel of magnets? 'To find out whether taking images of the male and female genitals during coitus is feasible, and whether former and current ideas about the anatomy during sexual intercourse … are based on assumptions or on facts,' Van Andel explained in his paper. The pursuit of truth knows no bounds.

But, of course, the Dutch scientists faced a seemingly insurmountable challenge: where would they find a couple who could perform the act in such a claustrophobic container, including a man who could keep it up under such conditions? This was 1999: Viagra had not yet hit the scene.

The solution: acrobats. One of the study co-authors, anthropologist Ida Sabelis, had previously cavorted as a street performer, and so had her boyfriend. The pair – lithe, flexible and not prone to stage fright – managed to fit inside the scanner (just 35 centimetres high) and perform the act of love in the name of science.

The results were published in 1999 in the prestigious *British Medical Journal* as, 'Magnetic Resonance Imaging of Male and Female Genitals During Coitus and Female Sexual Arousal'.[10] Conclusion: the penis does not drive into a woman like an impaling stick, but responsively bends at a 90 degree angle. Like a rubber boomerang.

The Dutch researchers won the 2000 Ig Nobel Prize in medicine, the feisty and funny counterpart to the mainstream scientific crowns, awarded to studies that 'first make people laugh, then make them think'.

Tellingly, Van Andel and his co-authors not only took naughty pictures, they also made a saucy video.* The biologists asked the technicians to stitch together single images into a film, like animators constructing cartoons from individual frames. But they did not release the video to the public for a decade, worried it might be condemned as pornographic and frivolous. When the film finally was unleashed upon the Internet in 2009, it became

---

* http://www.youtube.com/watch?v=OVAdCKaU3vY

a viral smash hit: as I write, it has been viewed more than 2.7 million times.

MRI studies have also revealed surprising features in the size and shape of the clitoris and how it changes when aroused. The visible bean is just the tip of the sensory iceberg. When the clitoral subsurface structures are made visible with magnets, dramatic changes can be seen: it doubles in size as it engorges with blood.[11] More on the subterranean landscape of the clitoris shortly.

The penis is not the only portion of the human sexual form that is distinct. Comparisons with the rest of the animal kingdom prove what adolescent boys suddenly realise: human breasts are special. They are of a unique construction, and no other animal bears structures like them. The udders of ruminants such as cows are formed of large empty sacs, and most other mammals have transient milk-producing glands that appear only during the period of nursing. We alone possess pendulous breasts that swell during adolescence* and remain enlarged for the rest of our lives, regardless of whether or not we reproduce. Our closest relatives – chimpanzees, bonobos, gorillas and orang-utans – have breasts that only swell during times of feeding.

Why should this be so? The scientific jury is out, but there are a couple of interesting ideas. Fans of the 'genital echo theory'[12] posit that we evolved permanent breasts to mimic the shape of our buttocks. Sociobiologists who endorse this theory point out that male monkeys are primed to look at the rumps of female monkeys to determine if they are in the fertile phase of their cycle. In many primates, the genitals of females swell to enormous

---

* Puberty results in biological changes that are so profound, both in the brain and in the body, some developmental biologists have posited that adolescence marks not only a weird and crucial turning point in our lives, but is in fact a form of metamorphosis.

sizes and intense colours when they are conceivable (technically termed 'coming into oestrus'). Many dog owners will recognise this, having seen the same physical transformation in their pets.

Male monkeys value the sight of female perinea so much that they will sacrifice portions of fruit juice in order to look at photographs (leading some to nickname this the 'monkey pay per view juice porn paper').[13] This same study found that male monkeys would also trade fruit juice to look at faces of high-ranking monkeys, but require the payment of fruit juice to look at images of the faces of low-ranking monkeys. Their conclusion: 'Monkeys differentially value the opportunity to acquire visual information about particular classes of social images.'

But when our ancestors shifted to walking upright, routine inspection of female genitalia was no longer easy from afar, so perhaps we evolved these unique structures to grab male attention. Maybe. It's an intriguing idea, but like all ideas from the realm of 'evolutionary psychology', which attempt to explain our behaviours today based on biological changes in eras past, it is pretty much impossible to prove.

Others assert that we have breasts for a simple but unsexy reason: fatty tissues provide energy reserves for women who in prehistoric times would have routinely faced the threat of periodic famine. You may find this explanation hard to believe considering what a source of obsession large breasts can be for some men and how much money so many women are willing to spend on silicone implants (even after the French PIP scandal, which saw more than 5,000 women suffer burst implants).

Moreover, as some of us anecdotally report, breasts act not only as visual attractors but spectacular triggers. Many women can climax to orgasm from nipple stimulation alone (and sometimes do so while nursing, a little discussed, embarrassing, but interesting quirk of our biology). By the by, third nipples are

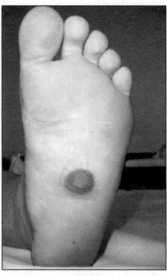

5. Six per cent of the human population has a supernumerary or 'third nipple'. Nifty and nimble, third nipples can be found just about anywhere on the body – even the foot.

more common than you might think: it is estimated that up to 1 in 18 men and 1 in 50 women – including Lily Allen, Zac Ephron and Mark Wahlberg (or Marky Mark as he'll always be known to my generation). Third nipples are impressive in their anatomical roving: they are habitually found all over the body, including the legs and the feet. The condition of having a third nipple, or polythelia, is however not as quirky as polymastia: bearing a third breast. Anne Boleyn, King Henry VIII's second wife, is said to have had one[14] (though this could simply be chalked up to political slander – we might never really know the truth). But today scientific reports have verified that third breasts do regularly crop up. Not just near the normal pair, but in strange and distant locations: a 2006 medical report described an Australian woman who developed one on her foot, complete with sweat glands.[15] Men too can develop third breasts; a 2001 study reported one unfortunate patient who developed a breast on his thigh, complete with a nipple.[16]

Nipples have other migratory manoeuvres. An illuminating 2011 study[17] found that – in the brain – they lie next to the clitoris, cervix and vagina in women (and next to the penis in men). The location of the nipples in the brain does not correspond to their corporeal location on the body: inside your head, they are not placed next to the torso or the neck, but next to the sex organs. The nipples snuggle alongside the genitals in a region of the brain called the 'somatosensory cortex': two strips which run along the left and right sides of the brain, running from just behind the ear up to the top of the head.

The 'somatosensory cortex' is a map of feeling: were I to poke the region of it that corresponds to your elbow, you would feel a tingling in the crook of your arm.

Thus in the brain there is a direct link between our nipples and our genitals. Or, in the scientific prose of the study authors: 'Activation of the genital sensory cortex by nipple self-stimulation was unexpected, but suggests a neurological basis for women's reports of its erotogenic quality.'

The mapping of these bits of the brain in the 1950s makes for an interesting story. When Canadian neuroscientist Wilder Penfield (1891–1976) and his contemporaries wanted to locate where anatomical regions are positioned in the brain, they used epileptics undergoing brain surgery. These patients suffered from frequent and incapacitating seizures so severe that they opted to have the tops of their skulls popped off and the tiny chunk of their brain responsible for their seizures removed.

The brain does not contain sensory nerve endings, meaning that if I popped the top off your head and poked your squishy grey matter, you wouldn't feel it. So brain surgery can be performed with local anaesthetic while the patient is awake.

As these epileptics lay fully conscious on the operating table, they consented to neurosurgeons rummaging around while in there. They kindly provided a unique opportunity for scientists to explore the brain in what were otherwise healthy study subjects.*

Direct probing of grey matter in the 1950s yielded remarkable maps of the brain. Neuroscientists discovered that just next to the somatosensory cortex, the map of feeling, runs a complimentary strip, the motor cortex, a map of motion. Were I to jab at the portion of your motor cortex that corresponds to your feet, your toes would involuntarily wiggle. As it happens, the feet lie directly next to the genitals in both the somatosensory and motor cortices – leading many to speculate that this helps explain the high frequency of the 'foot fetish'.

Working with the map of the somatosensory cortex, imaginative scientists in the 1950s produced a spectacularly ugly model of the human body that illustrates the relative importance of anatomical regions as reflected by the size of the somatosensory cortex devoted to each. The lips are massive, the tongue enormous, the torso and arms minuscule, and the hands gigantic.

We are endowed with enormous numbers of sensory nerve endings in our hands for obvious reasons. We use them to manipulate and comprehend our world. Conversely, we have very few nerve endings in our back. Have a friend place two fingers any distance apart on your spine and you'd be surprised at how hard it is to deduce the distance between them. The model was dubbed the 'homunculus' (Latin for 'little man'), named after a medieval belief that a tiny, slightly demonic human could be formed from sperm. (Recipe: Place semen in a flask and heat gently. After forty

---

* For this reason epileptics continue to serve as the medical bedrock for studying the building blocks of the mind.

6. The homunculus – Latin for 'little man' – is a staple of neuroscience textbooks, illustrating how much of your brain is apportioned to sensation in each anatomical portion.

days a tiny human form will develop. Feed with blood for another forty weeks. Result: homunculus.)

But here comes the plot twist: the scientists fibbed. The size of the somatosensory cortex devoted to the penis is gigantic, but researchers in the 1950s did not reflect this in their illustration. They deliberately depicted the phallus as being far smaller than it actually is in the brain. Museum plaster homunculus models continue to duplicate this omission. An updated version[18] grants the male member its due precedence.

The obvious joke of course follows that a man's penis is bigger in his mind than it is in reality – but the opposite is in fact true.

Tellingly, scientists did not bother to map the somatosensory locations of the cervix, nipples, vagina, and other sexual structures in women until 2011,[19] half a century later. But this time, neuroscientists did not have to carve open a woman's skull to

7. If only school textbooks reflected reality. This is how large
the penis would be if accurately represented in drawings that
illustrate the amount of brain space devoted to each body part.

poke at her brain: they were the lucky inheritors of modern
brain-imaging technologies. These allowed them to see which
portions of the brain 'lit up' using functional magnetic resonance
imaging (or fMRI), a form of brain scanning that uses magnets
to trace the flow of iron in blood as it moves through the body,
giving a rough idea of which parts of the brain are 'active'. While
women lay with their heads strapped into enormous claustrophobic scanners they followed instructions to stimulate themselves
in precise places. Result: the cervix, vagina, nipples and clitoris
all have distinct regions in the brain, underscoring the fact that
touching the clitoris is not the same as tugging the labia.

Instructing people to masturbate inside brain scanners has led
to intriguing insights about what is, for some, the penultimate
purpose of sex: orgasm. Studies have shown that the experience
of climax in women tickles up to 30 different regions of the brain,

including ancient structures such as the cerebellum (Latin for 'little brain', the wrinkly bulb at the back on the bottom, crucial for basic functions like maintaining balance); the nucleus accumbens (central to reward and pleasure, and the crucial point of manipulation for many drugs); and even the lower brainstem, a pointy tip at the base of the brain that plugs into the spinal cord (that most ancient part of our brains).[20]

Other studies[21] of orgasm in women have suggested that the prefrontal cortex – which sits directly behind the forehead and is associated with 'higher' cerebral functions like self-control, consciousness and rational thought – is 'deactivated' or 'switched off' during orgasm in women. This, of course, has led some to extrapolate from a glimmer in the neurological data to imply that women 'stop thinking' when they climax. Others have suggested orgasm is an 'altered' state of consciousness. Whatever it is, neuroscience has revealed that whatever happens in the brain at the moment the earth moves, it certainly is unlike anything else. Who knows what else scientists will unveil in the future?

We are not the only animals who enjoy the act, and we certainly are not the only ones with weird anatomies. But we are unique. More than any other species on earth, our bodies are built to produce pleasure from sex, and we probably experience more pleasure from sex than any other form of life that has ever existed. We have spent much of our time on this planet bewildered, intimidated and even angered by this unavoidable facet of our existence. Yet the truth is undeniable. Science has revealed sex for what it truly is: biologically unrivalled. It is a gift, and an integral component of the human condition. It played a crucial role in the evolutionary journey that made us who we are.

# 'SO PRETTY AND USEFUL A THING'

Vesalius – the 'father of anatomy' – did not know where the clitoris was.

When Italian Realdo Columbo (1516–1559) claimed to be the first to locate 'so pretty and useful a thing' in his seminal work *De Re Anatomica* (1559), Belgian anatomist Andries van Wezel (1514–1564), better known by his Latinised moniker 'Andreas Vesalius', in turn lambasted the 'discovery' of the organ in a letter to colleague Gabriele Falloppio (1523–1562), who also claimed to have 'discovered' the clitoris. Vesalius: 'You can hardly ascribe this new and useless part, as if it were an organ, to healthy women'.

Keywords: 'new' and 'useless'.

But Vesalius was not unique in his incapacity to locate this pretty and useful thing. The location and importance of the clitoris has been a matter of debate at numerous points in western history, well into the nineteenth century. The father of anatomy was capable of tracing the ducts that connect the umbilical cord to the fetus and illuminating the valves of the heart, yet he disputed the existence of the clitoris.

It contains more nerves than any other portion of the human body outside the brain. And the father of anatomy deemed the clitoris useless, irrelevant – and imaginary. It is amusing to note the difficulty that many people – men, scientists and, on occasion, even women – have in finding the clitoris, especially considering the ease many children have in locating their stimulatory appendage.

With more than 8,000 nerve endings, the human clitoris is unparalleled among mammalian morphologies. Biologists contest it is the only portion of the human form designed for pleasure alone. Male orgasmic ejaculation is compulsory for fertilisation to occur through the old-fashioned mode of reproduction, but female orgasm is not required for ovum to meet sperm. The clitoris is both ancillary and spectacular. Anatomists, evolutionary biologists, psychiatrists and the man on the street alike have all (let's hope) taken note that the clitoris is of biological significance.

That the very existence of such a remarkable organ could have merited any form of empirical investigation is striking, from both a cultural as well as scientific standpoint. How could anyone in possession of one, or partnered to someone with one, be unaware of its location or importance?

Ignorance of the clitoris is probably the exception rather than the norm for the human species across space and time. Most groups of people, spanning thousands of years, were probably more than familiar with the pretty and useful thing (certainly our cousins the bonobos are fans of the organ, stimulating the clitoris orally and manually, both heterosexually and homosexually).

Because the size and shape of the clitoris varies there have been numerous theories that clitoral volume correlates with personality traits: large clitorises were thought in the seventeenth and eighteenth centuries to be a sign of lesbian tendencies, and as such were termed 'tribade clitorises' (tribadism being another name for scissoring).

Medical historian Lesley Hall of the Wellcome Collection in London has curated an extremely lively collection of clitoral historical myths and facts,[22] including this quotation from Moris Farhi's 'Lentils in Paradise: A true and nostalgic account of my visits, as a little boy, to the Women's Baths in Ankara', in which he describes trips to the public pools where children of both sexes were admitted to bathe with naked women.

As for clitorises, it is common knowledge that, like penises, they vary in size. The Turks, so rooted in the land, had classified them into three distinct categories, naming each one after a popular food. Small clitorises were called 'susam', 'sesame'; 'mercimek', 'lentils' distinguished the medium sized ones – which, being in the majority, were also considered to be 'normal'; and 'nohut', 'chick-peas', identified those of large calibre. Women in possession of 'sesames' were invariably sullen; the smallness of their clitorises, though it seldom prevented them from enjoying sex to the full, inflicted upon them a ruthless sense of inferiority... Those endowed with 'chick-peas' were destined to ration their amorous activities since the abnormal size of their clitorises induced such intense pleasure that regular sex invariably damaged their hearts.

So while large clitorises throughout Europe were deemed a degenerate marker of homosexual predispositions, the Turks declared them a blessing and a gift.

Viennese psychiatrist Sigmund Freud (1856–1939), always the charmer, diagnosed them as childish. He declared the clitoris important for the achievement of orgasm, but only for immature girls who climax through masturbation. Through maturation and marriage, a woman would shift to climaxing vaginally rather than by direct manual or oral stimulation, a biological transition he dubbed the 'clitoral–vaginal transfer'. Most women, in Freud's opinion, would thus be developmentally stunted: survey results have varied, but the estimated percentage of women who do not experience orgasm from penetration alone typically ranges around 70 per cent.[23]

Contemporaries of Freud – psychiatrists and physicians throughout nineteenth-century Europe – are more famous for their manipulation of the clitoris as a means to cure female anxiety, depression, and other forms of psychological frustration. Physicians would apply continual stimulation to the vulva for up to a full hour before the patient would experience a 'hysterical paroxysm' – an orgasm or similar form of involuntary muscular response. We might snigger, but it's unlikely the women were having a good time. Modern masturbation measurements indicate the average woman can bring herself to orgasm in four minutes or less. These patients endured probing treatment for 60 minutes or more. Some argue the physicians weren't having a good time either, which led to the development of a range of electrified devices – proto-vibrators – that could execute the task more quickly.

The Victorian treatment was not unprecedented: the prescription of 'pelvic massage' as a treatment for 'hysteria' in fact dates back thousands of years before the Victorians adopted it. Manual medical treatments of the vulva were described in the works of Hippocrates, Celsus, Galen, and – linguistically most pleasing – Soranus. From Galen (translated by Rudolph Siegel):

> Following the warmth of the remedies and arising from the touch of the genital organs required by the treatment, there followed twitchings accompanied at the same time by pain and pleasure after which she emitted turbid and abundant sperm [this sperm most likely being female ejaculate]. From that time on she was free of all the evil she felt.

Hysteria, or 'womb fury', was most chivalrously described by Plato as 'an animal inside an animal'. The ancients ascribed anxiety, restlessness, sleeplessness and irritability to a 'wandering uterus'. This was thought to roam freely throughout the body, occasionally swaddling itself around a woman's windpipe.

The Italian physician Athonius Guaynerius prescribed the following treatment in 1440:[24]

Anoint the mouth of a vulva with different odoriferous materials, for which the prescription is also included, and to rub it into the neck of the womb as well. The rubbing, which should be done with the midwife's finger, will cause the womb to expel the sperm or corrupt humours and free the patient from disease.

We titter now, but let us not forget that hysteria was only removed from the Diagnostic and Statistical Manual of Mental Disorders (or DSM, the professional bible of psychiatric definitions) in 1980.

Today the new term for women deemed sexually maladapted in some fashion, 'female sexual dysfunction', could prove gangbusters for pharmaceutical companies seeking to profit off feminine anxiety with the same bonanza that Viagra produced.

Though not all modern medical treatments of female genitalia are motivated by money: Dr Pierre Foldès of the Saint Germain Poissy Hospital in France has developed ways to reconstruct the clitoris from the clitoral tissue that remains following female genital mutilation (FGM). More than 140 million women today are living victims of FGM, in which the clitoris, inner lips or both are removed from babies and children, almost invariably without anaesthetic or antiseptic.

In a recent study Dr Foldès describes how operations on 2,938 women (564 of whom had been mutilated in France, not Africa) between 1998 and 2009 resulted in improved sex lives for 81 per cent of his patients, and 29 per cent suffered less pain during sex. The kicker: 129 women began to experience orgasms for the very first time in their entire lives. Even women who have had their clitoris stolen from them can have it and its orgasmic power restored.

# COME AS YOU ARE

Whether or not or how a woman achieves an orgasm was and continues to be a source of anxiety, misery and joy– as well as a matter of scientific scrutiny.

American Alfred Kinsey (1894–1956), a lepidopterist turned sex specialist, studied the sexual behaviours and beliefs of the American public in the manner of an entomologist, meticulously cataloguing our quirky likes and dislikes in the manner of a butterfly collector archiving specimens. 'There's not much science here,' he is said to have remarked about the quality of mainstream medical beliefs concerning sex. He therefore sought to improve our understanding as any serious scientist would: by collecting data. Lots of it. In all, he and his team surveyed more than 18,000 Americans – but he had in fact hoped to interview more than 100,000 people.[25]

One of Kinsey's most incendiary findings: documenting that the prevalence of homosexuality was far higher than most Americans at the time assumed. He estimated that 37 per cent of men had experienced a homosexual encounter and 10 per cent of men are 'exclusively' homosexual. Kinsey disliked the parsing of people into exclusively 'gay' or 'straight' categories and declared that sexual preferences lie on a spectrum. Humans are not born into binary boxes.

'The study of sex was really almost a hidden subject at the beginning of the twentieth century – when Kinsey published his findings so publicly, the world went ballistic,' says Dame Anne Johnson,[26] Professor of Infectious Disease Epidemiology at University College London, who received a damehood in 2013 for her work on human sexuality and, in particular, the HIV epidemic in the 1980s. 'It was revolutionary that he legitimised sex as a legitimate topic for scientific inquiry.' Johnson is one of hundreds of scientists who have tirelessly catalogued British sexual habits with the National Survey of Sexual Attitudes and Lifestyles survey (Natsal) in the UK. Published once a decade (1990, 2000, and the most recent in 2013), these surveys have carried on Kinsey's meticulous mapping of human sexual diversity.[27]

Half a century ago, Kinsey's *Sexual Response in the Human Female* (1953), the follow-up to his ground-breaking study of the human male, *Sexual Behavior in the Human Male* (1948), indicated that 10 per cent of women had never reached orgasm in their marital intercourse, but that 40 per cent of women had their first orgasm through masturbation.[28] Kinsey would later conclude that climax through intercourse was the exception rather than the norm, and would pen passionate polemics railing against mainstream ideas that women ought to come through penetration alone.

Kinsey was married but was what would have been called at the time a 'latent homosexual'. This is unsurprising, considering one of his best-known studies measured the distance ejaculate travels. He aimed to address the question of whether or not the cervix actively sucks semen inwards to facilitate fertilisation, or if male emissions robustly slam against the cervix, forcing their way inwards. It seemed the answer could be found by measuring the velocity with which human ejaculate travels. To investigate, he paid more than 300 men to masturbate on film. Biographers have revealed that Kinsey's homosexual leanings resulted in much angst: he thrust a toothbrush down his urethra as a teen in an act of self-castigation, and later revealed faecophilic tendencies. He might have been a bit odd, but was undeniably a true humanitarian: he tirelessly fought for the notion that diversity is the norm and championed the importance of direct clitoral stimulation during a time when other post-war sex scientists trivialised it.

8. William Masters and Virginia Johnson. What could possibly look less creepy than enlisting your secretary to assist you in viewing randomly paired strangers having sex?

A case in point are researchers William Masters and Virginia Johnson – so famed they are known simply as 'Masters & Johnson'.

They took sex research beyond the meticulous categorisation Kinsey pioneered and brought it into the laboratory. They were the first to systematically observe couples copulating in the lab, in a sense revolutionising the study of reproduction. Defending their brave methods, they proclaimed: 'Science and scientists continue to be governed by fear – fear of public opinion, fear of religious intolerance, fear of political pressure, and above all, fear of bigotry and prejudice – as much within as without the professional world.'

However, the pair did many things that today would never pass the ethics committee, such as assigning random strangers to fornicate (modern anatomical studies enlist couples). The report was opaquely titled: 'Persons Studied in Pairs'.[29]

The two initially met in 1957 when Masters hired Johnson, his secretary, to assist him in his research, perhaps because he knew having a female author on his papers would lend his work a greater air of credibility. A man pairing random people to shag in front of him undoubtedly looks a bit suspect.

Masters and Johnson concluded that orgasms in women arise from clitoral stimulation, positing that tugs on the labia and clitoral sheath during penetrative sex deliver the requisite stimulation required for orgasm. Upside: the clitoris is crucial. However, they poisoned their clitoral chalice by arguing that women who do not climax from intercourse alone suffer from 'sexual dysfunction'.

Nor were they remotely unique in this designation: researchers Edmund Bergler and William S. Kroger defined frigidity as 'the incapacity of a woman to have a vaginal orgasm during intercourse' in their 1954 book *Kinsey's Myth of Female Sexuality: The Medical Facts*. Keyword: 'myth'. One can be certain that these two men were just *fantastic* in the sack.

Other researchers attempted to put forward more helpful suggestions for the possible use of the clitoris in penetrative intercourse. Surgeon W. G. Rathmann describes in a 1959 scientific paper[30] a clamp he devised to remove the clitoral hood to make it easier for fumbling men to find, claiming that one of his patients had 'wasted four perfectly good husbands' by not coming to him sooner.

And if hood removal sounds extreme, consider the extraordinary case of Princess Marie Bonaparte (1882–1962), who opted to have her clitoris surgically repositioned. Twice.

The great-grandniece of Napoleon Bonaparte, Marie was equal parts pioneering sex scientist as well as landmark medical case study. In her lifelong quest to experience orgasm through penetration alone, she sought to locate the anatomical source of her dissatisfaction. Hypothesis: The distance between the clitoris and the vagina predicts the likelihood of orgasm by penetration. Method: Interviews with 243 women, and measurement of their clitoral–vaginal disparity. A stickler for accuracy, she used the urethra as a point of measurement, rather than the vaginal opening itself, as the vagina's precise borders can be difficult to define.[31]

Conclusion: Women come* in three varieties. Luckiest: 'para-clitoridiennes', endowed with clitorises less than an inch divorced from their vagina, which she estimated at 69 per cent of the female population. Most unlucky: 'téléclitoridiennes', with more than an inch of distance to bear, estimated at 21 per cent. The rest: 'meso-clitoriennes'. Declaring herself to be an unlucky member of the téléclitoridienne, Princess Marie asked Viennese surgeon Josef Halban to loosen the earthly bonds of her clitoris by snipping the connecting ligaments and reposition her clitoris closer to her vagina. When the result proved unsatisfactory, she asked Halban to repeat the renovation. Again, to no avail for this woman who elevated climax through penetration alone to such a pinnacle that she underwent the knife – twice – for its achievement.†

The great game changer in the understanding of the demographics of female orgasm came with the famous *Hite Report*[32] in 1976 by American rogue researcher Shere Hite, who interviewed over 100,000 women, aged 14 to 78.

Her most frequently cited finding: 70 per cent of women do not experience orgasm through intercourse alone, but can do so routinely through masturbation, oral or manual stimulation of the clitoris. This figure continues to hold up as an estimate for

---

* Pun intended.

† Bonaparte's fate, however, is benign compared with that suffered by Lili Elbe (1882–1931), regarded as the world's first recipient of a male-to-female sex change. She died a year after her first operation (and a few months after her fifth).

female sexual response. Hite's findings were widely celebrated by women of all political stripes for demonstrating that climaxing solely from vaginal penetration is the exception rather than the norm. This brought psychological relief to an entire generation of insecure women (because every generation of women finds something to be insecure about).

It is interesting to note that Hite's methods were deemed by academics to be lacking in statistical rigour. Yet Masters & Johnson, who chose to study only women who climax from penetration alone (already established by Kinsey as a minority in the female population), did not endure equivalent criticism for their statistical sloppiness.

Masters & Johnson also recommended conversion therapy for the treatment of homosexuality, and made a good buck attempting to 'cure' gay men of their fondness for members of their own sex. Proving yet again (and this is a theme we will come back to) that scientists are people, people are flawed, and sometimes smart scientists not only hold asinine beliefs, but frequently do terrible things in the name of 'the truth'.

This is almost certainly more true of the study of sex than any other field in biology (certainly more than any topic in chemistry or physics). Sex is a powerful force, and a deeply personal one. Anything anyone in a position of intellectual authority has to say on the matter is therefore likely to hit a nerve. Researchers therefore frequently approach the field with an axe to grind. Their ideas have a profound influence on what we believe to be 'normal'. But their findings do not always lead to persecution and prejudice. Sometimes they deliver psychological relief and improve the lives of millions.

# 'THE TYRANNY OF THE CLITORIS'

For her contribution to our understanding of human sexuality, Professor Emerita Beverly Whipple was named one of the 50 greatest living scientists in the world by *New Scientist*. Dr Whipple (in partnership with her academic partner Dr John Perry) is responsible for coining the term 'G-spot'. They named this after German gynaecologist Dr Ernst Gräfenberg (1881–1957), who described the region – a nebulous clump of sensitive tissue in the ventral or tummy-side of the vagina, next to the urethra – in an academic paper in 1950.[33]

Over three decades, Whipple and Perry have studied the anatomy and physiological response of women to stimulation in the laboratory to defeat what they term the 'tyranny of the central role of the clitoris',[34] documenting through hundreds of papers the variety of orgasms that women experience: clitoral, vaginal (or G spot), uterine (through 'cervical jostling'), and blended (some combination of the above).

Though she will always be known as the woman who made famous the G-spot, Whipple has always stressed the importance of diversity: her work has examined 36 different regions that can be stimulated, from the ear lobes to the toes, and 15 different types of touch, from caressing and hugging to licking and tickling.

'When I give workshops, I stress that we will use a four letter word: Talk! Communication is the most important thing for enhancing intimacy and pleasure,' she says.[35] 'It's not about the right or the wrong way to experience pleasure. The best is not the enemy of the good. I hate the phrase "achieve orgasm". I prefer "experience orgasm". Sex should not be goal-oriented.'

But it is the G-spot that has garnered Whipple her fame. Other researchers have since taken up the cause, investigating,

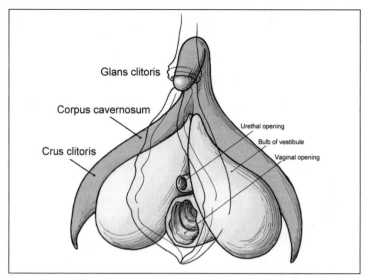

9. The clitoris: beneath the surface lies a branching
wishbone structure, the 'bulbs' of the organ. It
is far more than just a tiny nubbin.

documenting, proving and disproving the validity of the G-spot. Today its very existence remains a subject of scientific debate. As recently as 2012 a 'meta-analysis',[36] a study of studies published since 1950, concluded: 'Objective measures have failed to provide strong and consistent evidence for the existence of an anatomical site that could be related to the famed G-spot.'

The researchers, however, qualify this with this note: 'However, reliable reports and anecdotal testimonials of the existence of a highly sensitive area in the distal anterior vaginal wall raise the question of whether enough investigative modalities have been implemented in the search for the G-spot.'

Translation: loads of women report that having the tummy-side of their vagina stimulated in just the right way makes them come.

But another study that same year[37] – anatomical dissection of an 83-year-old cadaver – claimed to find proof of the organ, which the paper seductively describes as a 'well-delineated sac'.

The use of cadavers to map the anatomy of the female genitalia has also allowed scientists to reveal[38] that the clitoris is not merely a nub at the top of the vulva, but in fact a much larger, distinct structure, branching out into two wings that extend down around the mouth of the vagina like the breastbone of a bird. The Australian urologists who mapped them suggest they should be termed the 'bulbs of the clitoris'.

Another study in 2008[39] claimed to find anatomical proof of the G-spot with ultrasound: high-frequency sound waves, silent to our ears. This sound-based mapping, the Italian researchers say, could allow scientists to demarcate women who can orgasm through penetration alone from those who cannot. A *New Scientist* reporter airily divides these[40] into the 'lucky haves' and the 'have nots'. But anecdotal evidence indicates that a woman's propensity to climax from penetration alone changes over time, increasing with age. It is thought that the G-spot could have the capacity to grow and change, just like any other region of the human body, like a bicep or hamstring. The organ could be far more plastic than we realise.

Our collective cultural enshrinement of the importance, significance and superiority of the capacity to orgasm through penetration alone has resulted in the publication of countless scientific studies that purport to identify biological correlates, often reaching conclusions that are both asinine and insulting. Possibly the most hilarious is a supposed link between a woman's capacity to climax through penetration and the way she places her feet, simply titled: 'A woman's history of vaginal orgasm is discernible from her walk.'[41]

# 'DIE OF SHAME'

Beverly Whipple and John Perry were also the first to document in a modern scientific paper the phenomenon of female ejaculation.

The female capacity to expel profuse volumes of liquid has been known for thousands of years. The *Kama Sutra* refers to 'female semen'. Hippocrates noted the propulsion. In one of the more colourful descriptions, the seventeenth-century Dutch physician Reiner De Graaf (1641–1673) described in 1672 a 'pituito-serous juice' that makes 'women more libidinous with its pungency and saltiness and lubricates their sexual parts in agreeable fashion during coitus'. I enjoy his 'agreeable fashion' qualifier.

Many doubted that this release of fluid was a form of 'ejaculation', and instead countered that it could be something far less savoury: urination.

But other sex researchers were convinced the liquid was not urine, but an analogue to male ejaculation. In 1981, Whipple and Perry aimed to prove their hunch. They found a woman who could ejaculate easily and reliably, brought her into the laboratory, and analysed her expulsion.[42] An oft-repeated quote from Dr Whipple: 'On one observed occasion, the expulsion was of sufficient force to create a series of wet spots covering a distance of more than a metre.' The crucial bit of chemical detective work: the feminine excretion contained prostatic acid phosphatase (PSA), an enzyme found in male prostate secretions. They say this proves that the liquid is analogous to the liquid ingredients of semen. In fact, they think female ejaculation could have an evolutionary benefit: the liquid also contains very high levels of zinc, a potent antimicrobial. Women may have evolved the capacity to ejaculate in order to flush the urogenital system of dirt, microbes and infections.

Whipple and Perry devoted three decades to documenting the human orgasmic spectrum – ranging from women who have never had an orgasm, some 5 to 10 per cent of the population according to most estimates, to a woman who experienced 70 orgasms in the lab in an hour. Their description of her physiology is worth noting:

> One research subject masturbated with her own vibrator while her PC muscle [pubococcygeus muscle, the same that is strengthened with famous Kegel exercises] was being monitored by vaginal myography. During one hour we recorded some 70 discrete orgasms, consisting of 6 to 12 contractions of her PC in the form described by Masters & Johnson. The subject reported that at home she sometimes had 200 such orgasms in an hour – but that they were not very satisfying emotionally.

Crucially, Whipple and Perry did not conduct their work with the intention of lining us all along an axis of sexual performance. Rather, they hoped to provide us with information that would improve everyone's sex life: either because we can use their knowledge productively (such as learning to locate the G-spot) or feel assured that whatever our bodies do, it is nothing to feel crummy about.

Whipple and Perry's motivations have never been to spoil anyone's fun. For one, they demonstrated that women with spinal cord injuries and who are unable to walk can still experience orgasms through activation of the vagus nerve, which extends from the uterus and cervix to the brain (rather than the pudendal nerve, which extends into the genitals through the spine). Women who cannot walk can still climax.[43] They validated the existence of female ejaculation, which at that point in western culture was so little discussed, women who expelled during sex assumed they were urinating – and often sought butchering surgery to 'correct' their anatomy. Whipple and Perry recorded countless letters from thankful men and women who felt enormous relief in discovering they were not alone in their biological quirks, such as a woman who wanted to 'die of shame' every time she ejaculated.

Yet we as a species (at least in the anxious west) also use their research and that of all sex researchers to feel competitive with each other and bad about ourselves. Agony aunt and social psychologist Dr Petra Boynton of University College London says[44] what concerns her is that the way women (and men) experience orgasm isn't seen as a broad spectrum of normal responses, but as a form of achievement.

'The discussion of sex in the media – women's magazines and newspapers – has been narrowed down into an aspirational message: there is good sex out there that you should be having,' she says. 'The subtext is increasingly very judgemental, and there's a kind of cheerleading ghastliness to it all. We talk about sex in really odd ways – about positions and about products. But we don't talk about the conversations that need to take place between people. Everyone wonders if they are normal. And they are constantly told that they are not.'

Over the past two decades, women have increasingly agonised not only whether the way their genitals produce pleasure is 'normal', but also whether their physical appearance is 'normal'. Labiaplasty, surgical trimming of the labia minora, has gone mainstream, with thousands of women worldwide undergoing the knife to create what is colloquially known as a 'designer vagina'. Labia minora which have the audacity to protrude beyond the labia majora have been deemed undesirable.

Are the labia of these women unusually large? One of the very few studies that has looked at women undergoing labiaplasty from a statistical perspective found that 'the labia of all participants were within normal published limits'.

Pervasively, the data reveals that women who are perfectly average believe they are sexually abnormal. Yet all of us agonise over whether or not we are 'normal' – and this anxiety is not restricted to women.

## MADE TO MEASURE

Sexual agony and insecurity is not the sole preserve of women. Men too bear psychological anxieties. The most obvious source: penis size.

But does penis size *really* matter? It undoubtedly does, as evidenced by the sheer number of studies that have examined the size of the male member (as well as the importance that men attribute to it). The briefest of surveys online will throw up an enormous variety of papers examining the spectrum and significance of penis size. One of my favourites determines that contrary to popular belief shoe size has no relation to penis size.[45]

Kinsey the archivist systematically catalogued the variety in shape and size of the penis, asking his subjects to measure their length and report back with paper strips. An analysis of Kinsey's data 40 years later indicated that there were 'considerable discrepancies' between the actual length of the fleshy organs and the paper strips submitted: the men reported being larger than they actually were. Shocker.

Recent studies, based not on self-reported length but objective measurements by long-suffering measuring-tape bearing biologists (including one survey of 3,000 Italian men) show that men will underestimate the size of their penis in comparison to other men. A 2002 study[46] found that not a single one of 67 men seeking extension treatment were of a size (less than 1.6 inches flaccid) that would merit surgical alteration. 'Our data show that most men who seek penile lengthening surgery overestimate "normal" penile length,' the Florentine authors write. Tellingly, 62.7 per cent of the men said their insecurity began in childhood when they considered their length inadequate in comparison to their friends, and 37.3 per cent reported the problem began in adolescence when they began viewing pornography.

So, we know men think size matters. But surely the real barometer for the importance of penile volume ought to be the receptive perspective? Studies invariably show that while some women think it matters, most do not. And far more men than women think it matters.

One study[47] from 2006 found that while only 55 per cent of men were satisfied with the size of their penis, 85 per cent of women were satisfied with the size of their partner's penis. Another 55 per cent of women think the length of the penis is not important, and 22 per cent of all women said length is 'totally unimportant'. More than three-quarters of all women in this survey said they don't think length matters. Twenty per cent of women said length is 'important', and only 1 per cent said it is 'very important'. Other surveys have found variable results, but consistently show that women who think size matters are 'in the minority'.[48]

The simplest answer is that it depends on the woman – and on the man. Reducing the quality of a sexual encounter down to one single variable is hilarious at best, emotionally caustic at worst – and scientifically unsound.

Whether it is the shape of our bodies, how they produce pleasure, or what tickles the circuits in our minds, science continues to reveal that there is no such thing as 'normal'. Diversity is the norm.

# OF MONKEY BALLS AND MILLIONAIRES

It started with a simple recipe: remove the gonads of male dogs and hamsters. Grind, mix and inject. It led to the transplantation of the testicles of dead teenagers into the withered scrotums of ageing millionaires, and it ended with a farm in Italy devoted to harvesting monkey balls.

The testicle implant craze of the 1920s is one of science history's more bizarre chapters, but it proved integral to the development of one of biology's most complex and fascinating fields: endocrinology, the study of hormones.

Pioneer Mauritian Charles-Édouard Brown-Séquard (1817–1874) earned himself a place in medical history by popularising the idea that chemicals could be secreted by an organ and produce effects upon regions distant in the body. Today we know these as hormones: oestrogen and testosterone, oxytocin and adrenaline, the chemical messengers that produce shivers down the spine, racing in our hearts, and sweats upon our skin. Brown-Séquard's insights were groundbreaking. His most important achievement was identifying the adrenal glands – two pulpy caps that sit atop the kidneys – as the chemical factories of cortisol. In 1856 he demonstrated that this chemical is essential for life by removing the glands from lab animals, resulting in their death.

Cortisol is a trickster of a hormone: without it we would die, but too much of it can be disastrous. Nicknamed the 'stress hormone', it is released when we are aggravated, switching the body from the 'rest and digest' state to 'fight or flight'. Blood sugar spikes as fat and muscle are ransacked for energy. Heart rate soars. All physical reserves are recruited for immediate survival. Vital biological functions are temporarily marginalised. Cortisol impairs the immune system and the digestive tract to make way for muscular responsiveness.

This comes with severe downsides: prolonged elevation of cortisol – which frequently occurs in times of grief or stress – can lead to an enormous range of agonies, including depression, anxiety, chronic pain, autoimmune disorders, obesity and memory loss. Anyone who has suffered the ravages of post-traumatic stress disorder (such as myself) will attest that this natural hormone is as poisonous as any synthetic drug concocted in the lab. It can result in sleep deprivation, hallucinations, bone loss, thumping heart rates, and worse.

Yet without this molecular maverick we would die.

Such is the nature of hormones, the complex, irreplaceable messengers that travel throughout our bodies and allow distant organs to speak to each other. They are potent, and the intricacy of their networks is maddeningly complex. Unravelling their pathways today is not easy, and were even more difficult a century ago. Simply proving their biological significance was a challenge. And the strange experiments of the testicle implant fad played a part in demonstrating the importance of these invaluable chemicals.

In his later years Brown-Séquard, like many men, felt his vigour and energy fade. So he decided on a course of action in the fashion of an enterprising empiricist: he ate monkey testicle extract, and reported 'rejuvenated sexual prowess'. Next a ground mixture of dog and hamster testicles, semen, and blood was delivered by injection for the deliverance of prolonged life, heightened sexuality, sharpened cognition, and 'significant improvement in the urinary stream and the power of defecation'.[49] This *Brown–Séquard Elixir* was widely poo-pooed by his contemporaries.

But not all of them. The idea of ingesting, injecting or transplanting testicular tissue from a lively animal into a withering one seemed to make instinctive sense (tiger penis soup, after all, continues to appeal to those in search of a virile pick-me-up). Greek physician Skevos Zervos (1875–1966), inspired by the Ottoman vizier Kamil Pasha's taste for testicle soup, began experimenting with surgical rather than culinary solutions, transplanting the gonads of younger rabbits and dogs into older ones. After all, why bother with repeated injections of hormones if you can simply transplant the chemical factory that produces them, leaving the fleshy facilities to carry on manufacturing their wares? In 1910 Zervos took the next logical step: he implanted the testicle of an ape into a man, proclaiming his gonad transplant a miracle cure for both impotence and senility.

Word spread, and the idea gained traction. American G. Frank Lydston (1858–1923) published a paper[50] in the esteemed *Journal of the American Medical Association* in 1916 describing the benefits of transplanting the testicles of healthy young men (two teenagers who had died of accidental causes) into older men. Lydston believed so fully in the procedure that he had the operation performed on himself (as the recipient, not the donor, mind you). Dr Leo Stanley (1886–1976) was fortunate enough to have access to an even larger and more conveniently located population of healthy dead young men: executed prisoners. Stanley, chief surgeon at the San Quentin State Prison, transplanted testicles from 20 dead young men into older inmates, and relocated the gonads from deer, goats and other wildlife into more than 600 lifers. He describes his operations – which involved more than 1,000 men in all – in a 1922 scientific paper, frankly titled: 'An analysis of one thousand testicular substance implants'.[51] He claimed his glandular relocations could improve epilepsy, asthma and acne.

The true rock star of testicle implants, however, was Serge Voronoff (1866–1951), a Russian émigré to France, who first grafted a 2-centimetre-sized slice of chimpanzee testicle into a

10. Serge Voronoff, one of the world's first celebrity plastic surgeons, became internationally famous for transplanting slices of testicle tissue from monkeys into men – including himself.

man in 1920. His methods were intended to be more efficient and economical: why implant an entire organ into one patient when single slices will do?

Over the next five years Voronoff earned fame and fortune layering slices of monkey testicles into the withered sacks of well-moneyed aged customers. His experiments earned him such notoriety that E. E. Cummings immortalised him as a 'famous doctor who inserts monkey glands in millionaires'. Irving Berlin's 'Monkey Doodle-Doo' – featured in the Marx Brothers film *The Cocoanuts* (1929) – contains the convincing lyric, 'If you're too old for dancing / Get yourself a monkey gland'. Like the plastic surgeons of today, Voronoff enjoyed a plush life in Paris. When the French government banned the hunting of monkeys in the French colonies in Africa, he founded a farm in Italy for the rearing of healthy apes and harvesting of their cash crop. A gallery of before-and-after shots of his clients splashes across the pages of Voronoff's 1925 book *Rejuvenation by Grafting*, akin to the plastic surgery advertisements of today.

Parisian bars served drinks dubbed the 'Monkey Gland': gin, orange juice, grenadine and absinthe (which, to be honest, sounds a bit disgusting). Souvenir ashtrays bearing cartoons of startled monkeys hooting 'No Voronoff, you won't get me!' flew off the shelves. As late as the 1940s, football players at Wolverhampton Wanderers and Portsmouth were receiving treatments based on Voronoff's ideas.

Stateside, Americans took testicular surgery to even greater commercial success (as the American medical industry does), in the form of John Romulus Brinkley (1885–1942), a Kansas doctor who broadcast his services on his own radio station with the catchy slogan, 'A man is as old as his glands'. Brinkley was not in fact a doctor: he paid the Eclectic Medical University in Kansas $500 for his 'degree' – a sound investment, as he later charged $750 per surgery (equivalent to more than $10,000 today). He earned an estimated $12 million[52] by grafting goat testicles into gullible but desperate young American men who fell for Brinkley's claim that his treatments were best suited to intelligent men, but least suitable for the dim-witted. Clever.

But when Brinkley sued a rival physician for accusing him of charlatanism, the Texan jury voted in the rival's favour, declaring that Brinkley 'should be considered a charlatan and a quack in the ordinary, well-understood meaning of those words'. Voronoff too wound up publicly disgraced – even 40 years after his death. In the 1990s, an idea spread that his chimpanzee sex organ transplants could have been the ultimate source of the AIDS epidemic (though scientific analyses have found no evidence for this claim[53]). At best, the monkey slices most likely left little more than scar tissue: surgeon Dr David Hamilton in his 1986 book *The Monkey Gland Affair*[54] argues the transplanted tissue would be rejected by the host testicle, and leave only a scab. The perceived benefits would thus have been illusory.

Or … perhaps not? Subsequent research over the following decades has revealed that the Sertoli cells of the testicles – which form the walls of the sperm nursery, allowing for the safe growth and maturation of cells as they develop from spermatogonia into mature sperm – protect nascent sperm from the rest of the immune system. One could think of them as ejaculatory crucibles.

The testicles are just one of a palmful of fortunate body parts that the immune system will not attack if transplanted, along with the eyes, placenta and fetus.

Normally, when any one of us receives an organ transplant, such as a kidney or a lung, our bodies identify the new object as 'foreign'. It is as though the donor kidney simply wishes to wash our floors, but the trigger-happy border guards of our immune system block their entry for fear of invasion. We are thus required to take drugs that suppress the immune system from ungratefully rejecting the translocated organ and preventing it from taking up beneficial residence in our bodies.

The testicles are thus said to have 'immune privilege', and a 2005 Mexican study[55] made use of this: researchers coated pancreatic pig donor cells in human Sertoli cells before transplanting the porcine pancreas tissue into eleven diabetic children. None of the recipients required immunosuppressant drugs. It was a small study (and an ethically questionable one in the eyes of some[56]), but it seems that the testicle transplants of the 1920s may not have been as ineffective as they once seemed. Esteemed British journal *The Lancet* in 1991 officially called for the Medical Research Council to reopen the books and fund new research on monkey gland transplants.[57]

The work of the gland grafters thus serves as more than just a quirky chapter in the history of science. Their outlandish ideas and ballsy experiments laid the groundwork for today's life-saving xenotransplants, or inter-species organ relocations: pig heart transplants and suchlike. Human organ donors are always in short supply, and medical researchers have frequently looked to the animal kingdom for alternatives. Voronoff's work may yet prove to have had more weight behind it than we now realise.

That many of Voronoff's patients declared his surgical interventions a success speaks to the potency of the placebo effect and the power of the mind to influence the body. The science of sex has repeatedly demonstrated that sexuality is a psychological phenomenon as much as a physical one.

## PILLS AND POTIONS

Of course, all the psychological desire in the world will not help if the equipment refuses to work: impotence has been a source of misery for millions. Until the introduction of Viagra a colourful variety of potions, contraptions and implants were deployed in the war on unresponsive phalluses. Testicle soup, constrictive rings, testosterone injections, and eventually pneumatic devices have all played their part in the quest to get it up and keep it up. One of the oddest surely must be a cock ring formed from the eyelid of a sheep, with the eyelashes serving as a French tickler.[58]

But the real game changer came with the chemical sildenafil citrate, $C_{22}H_{30}N_6O_4S$, also known as Viagra, which works not by flooding the penis with testosterone,* as endocrinologists in the

---

* A friend of mine has Klinefelter syndrome: in addition to having an extra female chromosome, his body lacks the ability to produce normal levels of

early twentieth century had presumed would induce an erection, but by preventing the escape of blood. Sildenafil citrate blocks the enzyme phosphodiesterase type 5 (PDE5), causing the muscles to relax and for the spongy cavities identified by Leonardo to fill with blood. Thus Viagra and its counterparts are known as 'PDE5-inhibitors'. Inhibition leads to erection.

Serendipitously, sildenafil citrate was first given to men in clinical trials not in the search for a cure for impotence, but as a treatment for jet lag (due to the chemical's action on blood vessels, it is also used in the treatment of altitude sickness). When the study subjects proved reluctant to relinquish their potential jet lag cures at the trial's conclusion, researchers cottoned on – and the rest is history. Amusing the story is, but more importantly, it reveals that scientific progress is often far more facilitated by accident and luck, rather than by the targeted strikes that many researchers would have us believe drive progress.

Conversely, the hunt for the 'pink Viagra' – a treatment for the lack of arousal in women – has proved fruitless: no pharmaceutical drug has yet proven effective at enhancing a woman's likelihood of climaxing.* Should a pill that produces orgasm in women be found, the profits would no doubt be astronomical, as the history of Viagra and the vibrator demonstrate. Public awareness of 'female sexual dysfunction' was made possible by the publication of a now notorious paper in the supposedly esteemed *Journal of the American Medical Association* in 1999,[59] which purported to show that 43 per cent of women suffer from FSD.

---

testosterone for a boy. As a teen, he began receiving injections. This gave him a unique insight into exactly what it is that testosterone does, because he knows what it means to live without it. Simply put: 'It is the desire to try.'
* Though of course, alcohol, cannabis, MDMA and GHB are all said anecdotally to do the trick. Not being patentable – or legal – clinical studies have yet to be undertaken.

More than 40 per cent of women are sexually dysfunctional, you might rightly (and indignantly) ask? The reason for the finding is that the study authors defined FSD as having experienced pain or a lack of desire at *any* point in the past *year*, a 'statistical sleight of hand' as British sexual sociologist (and former sex worker) Dr Brooke Magniati puts it in her 2012 book *The Sex Myth: Why Everything We're Told Is Wrong.*[60]

'Characterising natural variations in women's libidos as a problem that needs to be solved is nothing new – [but] what is novel is the interest from big pharma companies in getting involved on a commercial scale,' she writes.

While industrial interest in FSD has continually grown in the past 14 years, scientists have continued to question whether FSD even exists – or what to call it. Other suggested terms under the umbrella of FSD include FSAD (female sexual arousal disorder), the official term of psychiatry's bible the DSM, FOD (female orgasmic disorder), and HSDD (hypoactive sexual desire disorder).* Just as Viagra illustrates neatly that scientific developments can be fortuitously furthered by chance, FSD exemplifies how medical progress – and our sexual well-being – can be manipulated by commercial interests.

## CHACUN À SON GOÛT

All things considered, the basic biology of human sexuality is pretty damn spectacular. Some may denigrate unadorned sexual preferences as 'vanilla', but scientific studies have taught us that

---

* Author Mary Roach suggests researchers may suffer from HAFD, hyperactive acronym formation disorder, in her delightful 2008 book *Bonk*.

plain old regular sex is, from an evolutionary point of view, unique. It is therefore surprising that there are so many of us for whom plain skin-on-skin is not sufficient to get our temperatures rising.

The spectrum of human fetishes – scientifically termed 'paraphilias' – spans the ridiculous to the downright revolting. Many will already be familiar with faecophilia (though my Microsoft Word dictionary is not), the sexual fetish for faeces, one of the most infamous of distasteful sexual deviances.

Another well-known sexual predilection is for our distant relatives: animals. Sheep shagging and enticing pets to perform oral sex on their owners by coating the genitals in peanut butter or chocolate are probably a common resort for those who lack a convenient human partner. Kinsey in fact estimated, based on his vast surveys of the habits of the American public, that 17 per cent of men who had worked on farms had experienced sexual contact with animals to the point of orgasm.[61] As the saying goes, 'there's not much to do in the countryside'.

Joking aside, sexual trysts with horses can be far from funny, as the very sad death of Kenneth Pinyan illustrates. Pinyan, an engineer with aeronautical firm Boeing, died in 2005 due to a perforated colon incurred through receptive anal sex with a horse. Male members of the species *Equus ferus caballus*, as you probably know, have large members, roughly 40 inches long when erect.

While legal frameworks prohibiting anal sex between two men were only lifted in all American states in 2003, new laws have recently been put in place to prevent people having sex with well-endowed animals.

Other well-documented paraphilias include: homeovestism, a weakness for people who 'dress appropriately'; formicophilia, the desire to be crawled on by insects (formicophilics frequently apply ants, cockroaches, and other crawling arthropods directly to their genitals); oculolinctus, licking the eyeballs (ouch); macrophilia, lust for giants; lactophilia, for breast milk; nasophilia, for noses; and paraphilic infantilism, also known as 'adult baby syndrome', which involves wearing a diaper and being cuddled like an infant. A particularly curious one: 'forniphilia', the fetish for transforming a lover into a piece of furniture. Members of this clade photograph their partners substituting for tables, functioning as bookshelves, and sporting lampshades.

There are some things described as 'paraphilias' that some of us might call 'pathologies', such as 'somnophilia', the fetish for sex with sleeping people (sounds like rape to me). 'Vorarephilia' is another: its enthusiasts get off on the idea of eating or being eaten by others.

The human brain is the most complex thing in the universe, and this intricacy bestows on it the capacity to find just about anything sexually exhilarating. Even if it's as innocuous as a lampshade, or as lethal as cannibalism.

## GROSS INDECENCY

At what point does a sexual predilection become pathological? Designations vary, and a great deal depends on definitions – and not just for academic reasons. Let us not forget that homosexuality was defined as a 'sociopathic personality disturbance' in the DSM until 1973. Happiness depends on what we define as 'normal', and countless lives have been rendered unliveable by the judgements and norms of the day.

For much of the nineteenth century, homosexuality was regarded neither as a choice nor a biologically predestined condition, but as a moral failing. In fact, for much of human history, homosexuality was not regarded as a 'condition' at all. Men who chose to have sex with men were considered guilty of committing buggery or sodomy, but not actually as being of a separate sort of man compared with their exclusively heterosexual counterparts. The word 'homosexual' only entered the English lexicon in 1869.

A notable case is that of pioneering computer scientist Alan Turing (1912–1954), who was convicted of 'gross indecency' in the United Kingdom in 1952 under an antiquated law dating back to 1885 and subjected to hormonal treatments to 'cure' his sexual attraction to men. Injections of the synthetic oestrogen Stilboestrol led to impotence and mammary growths. Turing committed suicide with cyanide two years later. His work was integral to the efforts to crack German encryption codes during the Second World War. And we thanked him with chemical castration.

'Homosexuality' was not officially removed from the DSM until 1973, and until this was achieved, gay men and women were frequently incarcerated in mental institutions. Subject to 'conversion therapy', ice bath treatments, debilitating pharmaceutical interventions, and even 'ice pick' lobotomy (which involved shoving a tool under the eyelid into the brain). More on this treatment later.

None of these actually 'cured' individuals of their 'pathology'.

Following from Kinsey's watershed studies showing the prevalence of homosexuality in the American public, combined with social movements to declassify homosexuality from psychiatric textbooks, several scientific milestones helped change western medical views.

One, the statistical analyses of Evelyn Hooker (1907–1996), a young female psychologist who subjected 30 gay and 30 straight men to a battery of standardised tests (including the

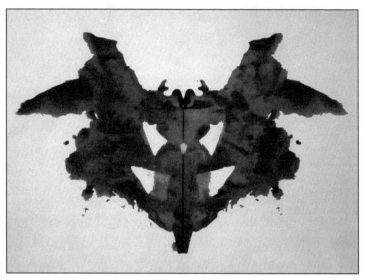

11. Once a staple of psychiatric diagnostics, the Rorschach
test was once deemed so reliable it was thought shrinks could
even detect homosexual leanings by a subject's visions.

famous Rorschach ink–blot test). Though she had been trained
to identify homosexuality as pathological, her friendships with
gay men led her to question the status quo. To avoid letting her
own predispositions skew the study, she gave the data to other
psychologists to analyse. Though senior academics were certain
they could recognise the difference between gay and straight men
(even based on what they saw in squiggly black ink swirls), the
results showed otherwise: there were no discernible psychiatric
differences between the two groups. When Hooker presented her
findings to the American Psychological Association in 1956, the
results did not go unnoticed.

But what caused more of a stir – simply by virtue of the
visual effect – were the public statements made by psychiatrist
Dr E. Fryer, a gay shrink unable to come out of the closet due to
stigma among his colleagues. In order to publicly speak about his

12. Gay shrink Dr E Fryer speaking in disguise for
fear of being professionally ostracised at the 1972
American Psychiatric Association's AGM.

professional barriers, he addressed the 1972 American Psychiatric Association's annual meeting under the name 'Dr H. Anonymous' – wearing a tuxedo, wig and rubber mask.

It worked. Homosexuality was removed from the world's psychiatric bible a year later.

But one question remained: *why* are people gay? Why any individual should prefer to mate with members of its own gender, with whom it cannot reproduce, is counter-intuitive. From an evolutionary standpoint, it's a puzzle. If evolutionary change involves the passing of traits to the next generation, shouldn't any gene that renders its owner unwilling to reproduce vanish from the collective inheritance pool? For this reason some biologists then (and even now) dubbed 'deleterious' any gene that could result in homosexual preferences.

Because the possibility that homosexuality could be genetically programmed seemed so improbable, many assumed the root cause of gay leanings to be 'environmental': a cold mother, a sexually abusive relative, and so on. Being gay was still a matter of will.

'This was tied to notions of free will, so of course there are important implications regarding choice – and obedience,' says Dr Qazi Rahman, a psychologist at the Institute of Psychiatry, King's College London. 'Biology matters when you use the immutability principle to deny gay rights. People used to think that sexual orientation was something you could learn, and that ended up influencing our ability to adapt.' The question is therefore important: are some people attracted to members of their own sex for reasons that are biological or environmental? In other words: nature or nurture?

Increasingly, evidence began to accumulate that suggested homosexuality could be genetically programmed. In 1963 a Yale biologist, creating new strains of the fruit fly *Drosophila melanogaster*, discovered a strain[62] in which males copulated with other males. He called the variant 'fruitless'. This posed an alluring question: could homosexuality in humans too be tied to genetics? Could there even be a 'gay gene' that programmes our preferences?

Genes are essentially the words that make up the human cookbook. Each gene is a portion of text in your DNA that codes for the manufacture of a protein, which themselves are the building blocks of every biological structure. The receptors in your eyes that receive light, the iron-containing globules in your blood that extract oxygen from the air in your lungs, the fatty lining that insulates the wires in your brain. All are made of protein, and all are coded for by genes. The human genome is a library of protein instruction manuals.

That one gene can result in one characteristic is clear: the genes for eye colour, colour-blindness, variants of deafness, and

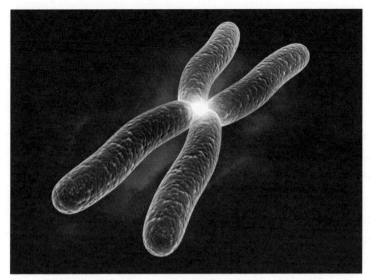

13. If the genome is a library of instruction manuals, you can think of a 'chromosome' as one book. There are 46 books in the library. In men and women 45 books are the same, except for the last: there is a book called X, and a book called Y. Men are XY. Women XX.

a huge number of other physical characteristics have all been identified.

But could genes really be responsible for something as complex as behaviour? The idea of a 'gay gene' didn't hold much traction for biologists and activists alike. But pieces of evidence that homosexual preferences are at least partially programmed by biology began to pile up. A study of identical male twins in 1991 indicated that if one twin is gay, the other is three times more likely to be homosexual too than if the twins are non-identical.[63] Two years later another biologist purported to have discovered a region on the human X chromosome which they said predisposed men to homosexuality,[64] dubbing the gene 'GAY-1' (slightly reminiscent of graffiti on a WC wall).

If the genome is a library of instruction manuals, you can think of a 'chromosome' as one book. There are 46 books in the library. In men and women 45 are the same, except for the last: There is a book called X, and a book called Y: men are XY, women are XX.

That is the main genetic difference between our genders: one tome in a library of 46 books.

More recent genetic studies from Italy have found that the female relatives of gay men have increased fertility,[65] claiming to have found a causal explanation for how homosexuality could exist at all. Such women would thus carry a gene that makes their sons more likely to be gay, but their straight daughters more fertile.

In the 1990s, anatomical evidence began to appear. A small part of the hypothalamus – an ancient portion of the brain crucial to basic drives like hunger, fatigue and sexual desire – called the 'third interstitial nucleus of the anterior hypothalamus' (INAH3) appeared to be the same size in straight women and gay men. But this region varies in size among gay men themselves – by as much as 20 times – and the initial study was conducted through autopsies on gay victims of AIDS, whose overall physical deterioration implies they may not have been representative of gay men as a whole. So the merit of the finding remains questionable. Still, other neuroscientists in the early 1990s found small differences in the brains of gay and straight men: the 'anterior commissure' – a bundle of fibres connecting the left and right parts of the brain – seemed to be 34 per cent larger in gay men than straight.[66] The 'suprachiasmatic nucleus' appeared larger among homosexuals.[67] The SCN is normally involved in keeping time, so nobody is really sure what this means (if it is relevant at all).

Inspired by the potential to find genetic roots for homosexual predispositions, biologists searched for other anatomical clues.

Fingerprints among gay men appear to be more similar to those of straight women than straight men. The difference in size of the index finger and the ring finger – the '2D : 4D' ratio – was also deemed a mark of homosexuality[68] (a small 2D : 4D ratio means the index finger is much shorter than the ring finger). Gay men seem more often to have index fingers that are longer than their ring finger, and it is thought that this is linked to levels of testosterone in the uterus during pregnancy.

In other words, it could not just be our childhood that influences our sexual disposition, but environmental conditions in the womb itself. Studies on birth order have suggested that, in families with many boys, younger ones are more likely to be gay than older ones, a result known as the 'fraternal birth order effect'. Crunch the numbers, says chief proponent of the idea Professor Ray Blanchard of the University of Toronto,[69] and it breaks down like this: if you are a man, every older brother you have increases your chances of being gay by 33 per cent.

Could the cause simply be big brother bullying? Unlikely, says Blanchard. He and others believe that remnants of male fetal cells left behind in the womb and the mother's bloodstream provoke an escalating counterstrike. The mother's immune system produces more and more antibodies against male hormones, resulting in lower levels of testosterone in the uterus.[70] 'Low prenatal testosterone could foster male homosexuality and an unusually high level of androgens [male hormones] produced by the fetal adrenals in baby girls could foster female homosexuality,' explains Dr Roger Gorski of the University of California,[71] who has studied the influence of hormones on sexual predisposition for decades.

The nurturing plains of the uterine cavity are more of a battleground than we thought.

So are the causes of homosexuality environmental – or genetic? The answer is, as with almost anything in biology, both. Genes will behave differently under varied environmental conditions: it is the gene–environment interaction that ultimately matters.

Most importantly, studies of homosexuality illustrate one underappreciated aspect of the scientific process: science is not an inexorable march towards 'the truth'. Ideas and the evidence gathered to support them are formed by people, who live in the world, have their own biases, and thus ideas are always dependent to some degree on what scientists have personally experienced. Scientific evidence is far more influenced by our emotions than empirical idealists would have us believe.


## EVOLUTION'S RAINBOW

One of the most salient clues to the origins of homosexual leanings may come not from psychological studies of human subjects, but from a broader population: the animal kingdom.

Homosexuality is not a quirk of humanity: it is widespread. Research has revealed that animals across the evolutionary tree, from macaques to caribou, jewel fish to black-spotted frogs, have homosexual denizens. And their population size can be substantial: up to 31 per cent of albatross pairings are female–female.[72]

According to Professor Emeritus Joan Roughgarden of Stanford University,[73] homosexual behaviour in animals isn't an occasional biological oddity – it's the norm. She has spent decades documenting same-sex liaisons in fish, birds, reptiles and amphibians, all up and down the evolutionary tree. She makes a case for the idea that same-sex contact in animals demonstrates that it is not restricted to humans, and that the trait is not due to a deleterious gene, doomed to be whittled out of the collective gene pool.

Professor Roughgarden has not only documented homosexual

behaviour in animals for decades, she transitioned from male to female in 1998 at the age of 52. Needless to say, she has handled an enormous amount of scrutiny and criticism of her work and her personal life.

She has studied the sexual habits of animals for four decades, during which time her argument – that same-sex pairings are rife throughout the animal kingdom – has frequently been ridiculed or ignored. Homosexual behaviour, declared her critics, was pathological: a genetic aberrance that natural selection should never favour.

'In the decade since I published my book, I think biologists at last recognise the ubiquity of all sorts of expressions of gender and sexuality that don't coincide with the binary picture you find throughout biology,' she says.

Moreover, she says, most views that homosexuality in animals or humans derives from environmental circumstance – such as overpopulation – imply that homosexual contact is 'second best'.

'These ideas are condescending. My view is that homosexuality is actually a positive trait, and one that is adaptive,' she says. 'Homosexuality is a special case of animal intimacy, and that's the real issue: explaining intimacy in animals, which sometimes but doesn't always involve genital contact. Behaviours that involve physical intimacy are the key to understanding how animals cooperate. Intimacy is a mechanism for producing cooperative teamwork in a network of relationships – and intimacy is a concept that is usually owned by the humanities.'

In other words, love, attachment, or simply taking pleasure in another creature's pleasure – regardless of the reproductive outcome – are part of our biological heritage.

Many of the most academically influential and publicly well-known theories concerning evolution have focused on the combative and violent aspects of survival. The phrases 'red in tooth and claw' (Tennyson) or 'the selfish gene' (Richard Dawkins) stick in our collective consciousness for a reason. The evolutionary value in besting one's enemy seems to make instinctive sense. Yet increasingly biologists are examining how cooperative behaviour – from termites to moles and humans – drastically improve the odds of a species' survival.

And science increasingly reveals that sexual contact seems to be an important component of cooperation. In other words, sex is not just about reproduction: it is also about social navigation. We often speak of the 'war between the sexes', but love within the sexes is increasingly appreciated as relevant.

Intimate contact between members of the same sex might not just be an anomaly, but in fact, evolutionarily advantageous. It might possibly not just be part of what makes us human, but one of the things that made us such a successful and widespread species in the first place.

## INCONCEIVABLE...

Syphilis disfigures, HIV kills, HPV causes cancer, and pregnancy (depending on your perspective) changes your life forever. And yet still our population abounds with men who do not want to wear a latex condom, withering at the mere sight of one.

Prior to the development of vulcanised rubber by the American inventor Charles Goodyear (1800–1860) in 1844, previous incarnations of the condom were fashioned from far less subtle materials, including silk, oiled paper, leather, linen, and – yes really – tortoiseshell.[74] Even when rubber hit the scene, the early models of the nineteenth century were as thick 'as a bicycle inner tube', and sported a thick seam on one side. Perhaps even less appealing, many were not designed to fit around the entire penis, but as a cap to fit over the glans (but under the foreskin). Or inserted straight into the urethra.

The widely fabled 'Dr Condom' – supposedly the etymological origin of the device's name – almost certainly did not exist. The true inventor of the condom is generally accepted to be the Italian anatomist Gabriele Falloppio (1523–1562), whose many accomplishments include the elucidation of the Fallopian tubes (which lead from the ovaries to the uterus).

Birth control pills, latex condoms and rubber diaphragms have taken us a long way from the nascent origins of prophylactics. Previous eras too were distinguished by a great spectrum of devices, materials and methods that have been employed in the quest to enjoy the act of sex divorced from the reproductive outcome. Crocodile dung, pebbles, honey, lemon rinds, and mule earwax have all been deployed as a barrier method (layman's terms: up the vag). As a more drastic course, abortive tonics have also found favour, such as toxic mercury-laced cocktails, beaver testicle moonshine, and high-potency gin (hence the nickname 'mother's ruin'). More recently, douches – vinegar-laced vaginal sprays – were endorsed as a form of contraception long before they were advertised as a means to 'clean' female nether regions and rid women of the supposedly 'unpleasant' smells they were informed they possessed, leading to infections and worse.

Inevitably we come to the birth of one of the most historically important and biologically innovative contraceptive methods: the combined progesterone–oestrogen pill. It is often said that it changed the course of history, as well as making possible the heady days of 'free love' in the 1960s. While Viagra's power to enliven the male member was discovered purely by accident, the quest to produce a safe and reliable pharmaceutical contraceptive was made possible by sheer determination and political conviction.

The story begins with chemist Russell Marker (1902–1995), who sought to produce synthetic hormones, and progesterone in particular, which in 1938 was expensive to synthesise. Searching for a cheap substitute, his chemical excursions brought him to the world of saponins: botanical molecules that behave like animal steroids (a class of hormones that includes testosterone, oestrogen, progesterone and cholesterol).

His search brought him to Mexico and the wild yam species *Dioscorea mexicana* and *Dioscorea composite*, from which a compound called diosgenin can be extracted and converted into progesterone.

Marker was turned down by two different drug companies before he decided to found his own pharmaceutical firm Syntex in Mexico City in 1944.

Work by chemists at Syntex and elsewhere on synthetic hormones ticked along for a few years,* but progress proved frustratingly slow for women's rights activist Margaret Sanger (1879–1966) and scientifically educated philanthropist Katherine Dexter McCormick (1875–1967). Convinced that women needed a safe prophylactic they could control, Sanger searched for a capable chemist who could speed up the progress of research, and convinced wealthy McCormick to fund Marker's work. Result, the Pincus Progesterone Project, led by scientist Gregory Pincus (1903–1967). The trial was nicknamed PPP to reflect the considerable volumes of female urine required for analysis (pee pee pee).

Clinical trials initially carried out in the slums of American territory Puerto Rico with progesterone-only pills gave mixed results, until chemists removed 'impurities', and the results were

---

* Norethynodrel and norethisterone, which later formed components of the early birth control pills, were initially marketed as treatments for mental disturbance.

even shoddier. Those 'impurities' turned out to be oestrogen, and thus the solution was found: a pill combining oestrogen with progesterone. Again, demonstrating that scientific discoveries are often the result of serendipity and accident.

And in this case, political mettle and the determination to defy established norms played an integral role in furthering scientific understanding and achievement. 'When the history of our civilisation is written, it will be a biological history and Margaret Sanger will be its heroine,' correctly predicted H. G. Wells in 1931.

Thus, a tool for reproductive control was placed in the hands of women.

Though there may be a bias towards designing contraceptives that interfere with female rather than male biology, there are mechanistic reasons for choosing to barricade the cervix rather than stem the tide of sperm. Today the most popular form of contraception worldwide is not the birth control pill or the condom, as you might expect, but the IUD (intrauterine device). Interesting to note considering we still do not entirely understand how the 'coil', as it is nicknamed by the Brits, actually works.

It's generally accepted that the twisted metal devices 'interfere' with the environment of the uterus in some way, but exactly how is not known. Certainly our predecessors' habit of shoving foreign objects into the plains of the vaginal and uterine cavities would diminish fertility: the Maori of New Zealand are said to have shoved pebbles upwards for this very reason, as well as the ancient Egyptians, who introduced both fumigating gases and crocodile dung into the vaginal canal. Such is the power of sex: women are willing to insert faeces into their holiest of holies, and men have been equally willing to stick their member into it for the sake of enjoying copulation without fertilisation.

# INTOXICATING ODOURS

Smell is an almost indescribable force: that one scent that is uniquely irresistible. Though someone's face, hands, eyes and voice are also distinct, there is something about aroma that is individual, instantly arousing, and impossible to capture with words. There is no one-size-fits-all definition of what defines a sexy smell: what is intoxicating for some could be repellent to others. There is no barometer by which we could choose the Julia Roberts or George Clooney of scent.

'We humans usually think that we pick our mates according to how they look – we think of "love at first sight" – but we don't appreciate the importance of smell,' says Dr Leslie Knapp, who studies the genetics of immunity at the University of Utah (previously at the University of Cambridge) and a global authority on the relationship between smell and attraction.[75] 'But studies of primates and even studies of humans have shown that our ability to smell is very important, even in present-day society. How we perceive the smell of someone has an influence on how we react to them. There is good evidence to suggest it is an important factor in how we choose our mates.'

Napoleon is famously reputed to have written to Josephine: 'Will return to Paris tomorrow evening. Don't wash.' Canadian gay gospel outfit The Hidden Cameras croon in their track 'The Smell of Happiness', from the album *The Smell of Our Own* (2003): 'Happiness has a smell I inhale… I feed my own face when I soon crave a taste of the neck of a boy.'

Smell is our oldest sense: before our primordial ancestors could see, touch or hear, they could smell. Detection of water or airborne chemicals provided our very first sensory means to understand and navigate the world around us. So ancient is this sense, it is the only one that plugs directly into the cortex of the brain without first being mediated by the gatekeeper of the brain, the thalamus. The cortex is the outer shell, where other structures we have met such as the somatosensory cortex reside.

Neurons from the ear and eye first pass through the thalamus, considered the central switchboard for incoming sensory inputs before those signals are then relayed into the rest of the brain. But the highways of the nervous system that lead from the nose to the brain do so without bothering with the middleman of the thalamus: they penetrate directly into the olfactory cortex, hinting at an ancient role smell plays in our biology.

Yet we know less about this sense than any other. The Nobel Prize in physiology or medicine was awarded in 1967 for discoveries illuminating the mechanics of vision, and in 1961 for elucidation of the structure of the inner ear. But it was not until 2004 that work on olfaction would garner researchers equivalent recognition for revealing receptors in the nose. We still don't understand exactly how those receptors work, but we do know that the genes that code for odour receptors comprise one of the largest families of inherited units: more than 600 genes[76] in all programme the locks in our nose into which the keys of aroma fit (and possibly more – mice have more than 1,000).

So why would one person smell sweet to some and sour to others? Scientists, wielding the modern tools of genetic analysis, have started to unravel why and how our individual preferences are determined by subconscious cues. It turns out that the same genes that determine what we smell like also programme how our immune systems operate.

These genes together produce a collection called the 'major histocompatibility complex' (MHC). In humans they are known as 'human leucocyte antigens' (HLA), but they play the exact same role. MHC proteins cover the outside surface of white blood cells, and determine what kind of foreign invaders our bodies are able to identify and destroy.

Think of them as tiny molecular probes that bristle on the surface of a cell, constantly probing every intruder, asking the simple question: friend or foe? These molecular sentinels that staff the pulpits of your immune defences also design the shape of your individual odour.

In humans, there are 200 different genes[77] that code for MHC proteins (half of which are known to have immune functions). Each gene can come in a huge variety of forms. The MHC complex is the most diverse set of sensory genes known in vertebrates. There are millions of possible combinations, explaining why the spectrum of human scent is far more varied than hair or eye colour.

Each of us contains a different suite of MHC genes, and the greater the diversity, the more robust the immune system. Those lucky enough to be born with a broad swathe of MHC genes are better equipped to fight off disease because their immune systems have a more diverse and nimble army of soldiers manning the cellular membranes, and thus can identify a greater range of invaders. This is why inbreeding is a bad idea: if you procreate with a cousin or close relative, you are likely to have similar sets of MHC genes, and therefore could bear offspring with a depleted array of inherited immunities.

For this reason, the genes that code for MHC proteins are also known as your 'compatibility genes': not because they influence your romantic compatibility with a mate (though they seem to), but because they determine if an organ from a donor will be a good match.

Lab animals, from stickleback fish to rats and fruit flies will preferentially choose mates who possess MHC clusters that are dissimilar to their own. Dr Knapp's own research in mandrills and lemurs has demonstrated that individuals will use smell to 'identify potential partners with the appropriate genes,' as she puts it.

Does human olfaction operate in the same way? Do we sniff out mates by subconscious cues that protect against inbreeding and produce more robust immune systems in our sprogs?

The first study to provide concrete evidence for this idea came in 1995 at the University of Lausanne in Switzerland, where Professor Claus Wedekind used the dirty clothing of young men to probe the preferences of sharp-nosed women: 44 men slept in T-shirts for two nights, and 49 women were asked to rate how appealing they found all 49 odours. Wedekind had analysed the HLA genes of all 93 people. Indeed, he found women preferred the scents of men whose genes differed from their own.[78]

This inspired a flurry of research into the role of body odour in attraction. Scent seemed to offer an unexplored vista where little was known, and intrepid researchers could stamp their mark on the history of science. One consistent (and in the eyes of some, troubling) finding was that when women go on the contraceptive pill, their preferences change. They become attracted to the scent of individuals with MHC gene combinations that are more similar to their own, which biologists believe is a protective mechanism. Recall that the pill 'tricks' the body into thinking it is already pregnant; women 'with child' thus would benefit from preferring the company of their family rather than seeking out new sexual partners. Or so the argument goes – other biologists think pregnant women would benefit from promiscuous behaviour by tricking multiple men into thinking the child is theirs.

One intriguing study[79] found that if women were shown photographs of men and given a selection of smell samples from those same men (though without knowing which shirt belonged to which man), their visual choices frequently matched their olfactory ones: the scents they deemed sexy often came from faces they declared handsome. However, more recent studies have not replicated this result. Others came to the opposite conclusion, with women rating as attractive the faces of men with MHC genes similar to their own.[80]

Other researchers have examined the genetic compatibility of romantic pairs. One 2006 study[81] of 48 couples found that when women shared a higher number of MHC genes with their husbands, they were more likely to cheat. But not all studies have replicated those results. Position in the menstrual cycle, male 'dominance' status, and facial symmetry have all been scrutinised for links with MHC genetics and odour, and results for the past two decades have been conflicting (and confusing).[82]

Moreover, biologists still don't know exactly how MHC genes would actually influence body odour.

One of the reasons for the variability in research results, says Dr Jan Havlicek of Charles University in the Czech Republic,[83] is that there are so many other variables that can affect body odour: health, environmental pollutants, diet. He has found that men who eat meat are deemed less aromatically pleasing than vegetarians.[84]

'I am sometimes jealous of my colleagues who work with faces and visual cues. Smell research is difficult because it is affected by so many variables,' he says.

Could the key to our olfactory preferences lie in that holy grail for sensory researchers: pheromones? These chemical sirens are well known to act as sexual attractants in animals across the evolutionary spectrum, from fish to insects, birds and mammals. It seems perhaps obvious that pheromones therefore are at play in the hidden forces that make one particular person irresistible. Yet decades of chemical hunting have proved fruitless.

'What evidence is there that we don't have pheromones? None. But nobody has shown that they do or don't exist,' explains Dr Tristram Wyatt of the University of Oxford, a global authority on pheromones in animals.[85] 'I am confident, however, that they will be found: maybe not in my lifetime, but they will.'

Dr Havlicek for his part is less certain. 'I don't believe we will find compounds or mixtures of chemicals that make somebody irresistible,' he says. 'We might be affected by smell subconsciously, but that doesn't mean we are robots who just follow smells blindly.'

Yet it is undeniable that the way somebody smells can be uniquely seductive in a way few things can, says Dr Havlicek. And there is one avenue of research he thinks will be the most promising: bacteria. Not the germs our immune system fights, but the beneficial bacteria that live in, around and throughout our entire bodies.

Each one of us is home to 100 trillion bacteria. More than 10,000 species call the human species home, and each of us is home to a unique subset. About 1,000 strains of germ live in your body, from the innards of your ear to the crevices of your toes to the lining of your gut (which alone contains a full kilogram of bacteria). For every one 'you' cell, there are ten bacterial cells. You are 90 per cent bacteria. These are not microbial freeloaders: you need them just as much as they need you. Without them, you could not digest food, produce vitamin B, fight off nefarious viral invaders, or live at all.

On the surface of your skin, your body secretes sessile chemicals called 'androstenes'. These in turn are converted by bacteria into 'androstenols', winged molecules capable of becoming airborne and snuggling into the nostrils of a prospective bed-buddy. You smell sexy to that special somebody because of your germs.

We are just beginning to scratch the surface (in every way)

of how our microbial inhabitants make us who we are: following from the Human Genome Project, which mapped the dictionary of the human species, has come the Human Microbiome Project. This five-year endeavour, which ran from 2007 to 2012 and cost $195 million, discovered that there are in fact 8 million microbial genes in our microbiome – compared with a pithy 22,000 genes in the human genome. Microbiologists are only starting to mine the data and understand what this means. One thing is for certain: you are an ecosystem, and your tiny inhabitants have more to do with what makes you alluring than you ever imagined.

## FEAST OF STENCHES

I expected to smell better than two boys who had not washed for 40 days. I did not expect to be deemed less attractive than an orang-utan.

In 2011 Guerilla Science presented our audience at a music festival, the Secret Garden Party, with an array of human scents for them to sample, judge and rate: two boys, a woman (myself), and Hannah, a female orangutan, only revealed to be non-human after the judging.

More than 50 noses sniffed our 'smell stations', plastic boxes containing ripped shreds of fabric from T-shirts worn by our four research subjects.

To thicken the plot (or, at least, their odours), we had our two male subjects refrain from washing for 40 days and 40 nights in the run-up to the festival.

'This really was something that no-one had ever done before – this is a totally unique experiment,' says Dr Knapp. 'Humans

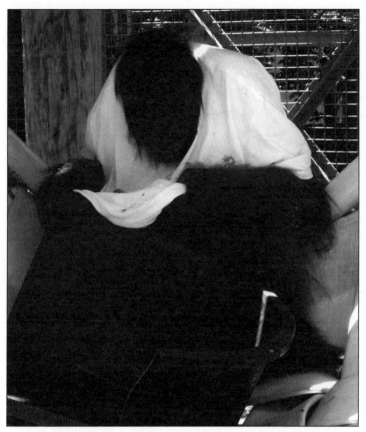

14. Hannah the Orangutan pitted her body odour against the smells of two filthy boys and one girl (clean, mind you) in a contest dubbed 'The Feast of Stenches'.

usually try to cover up their natural odours, so we were interested in the results.'

Our audience's first task was to guess the sex of each of the four stations using coloured beads to indicate male (blue) or female (pink). The overwhelming majority vote in every case was correct. Most people could tell that Daniel and Jim were male, and Hannah and myself female. Humiliatingly, more

15. 'Smell stations' contained ripped strips of fabric from the shirts
worn by the two dirty boys, the clean girl, and the hairy primate.
Truth be told: even handling the boys' shirts was unbearable.

people thought Hannah was female than deemed my scent
feminine.

We then asked the audience to rate the four smells on a sliding
scale of attractiveness. Daniel was deemed least attractive, Jim
less so, followed by myself. Hannah, the orang-utan, was declared
the most pleasing.

Our last table featured strips from four shirts, all worn by
Daniel, at various points during his 40-day soap fast. Unsurprisingly, the shirts worn at the latest stages were deemed unbearably
putrid.

But here's where things get more interesting: Dr Knapp
examined our DNA, extracted from our mouths by swabbing

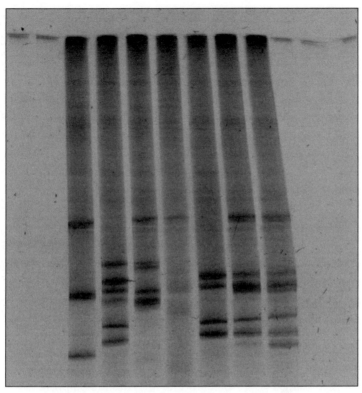

16. A 'gel electrophoresis assay', which illustrates the genetic qualities of a sample of DNA. In this case, the number of bands indicated the level of genetic variation, or richness, in each test subject.

the inner cheek, and produced visual illustrations (in the form of a 'gel electrophoresis assay', which displays genetic variations with bands).

She also examined the DNA of Shamima, Daniel's girlfriend at the time. Unsurprisingly, Sham (who descends from both European and Asian parents) boasted diverse HLA genetics; Daniel, whose lineage is largely Irish, possessed more homogenous genes, which might make him less immunologically robust and potentially more vulnerable to certain diseases than Sham.

But, amusingly, Sham deemed the smell of the other boy, Jim, to be more attractive than Daniel's – and Jim's genetics looked to be a better match if they were aiming for maximum HLA diversity, said Dr Knapp. Ouch.

## JUST ONE?

Few facets of the human sexual condition have been scrutinised, judged, maligned, ridiculed or canonised as monogamy.

Many biologists assert that monogamous pair-bonding is one of the few things that define our weird species. As famed primatologist and Director of the Living Links Center at the Yerkes National Primate Research Center Frans de Waal puts it:

> The intimate male–female relationship ... which zoologists have dubbed a 'pair-bond', is bred into our bones. I believe this is what sets us apart from the apes more than anything else.[86]

Biologists at Yerkes and elsewhere are investigating the genetics, neurochemistry and biological quirks of the little-known prairie vole *Microtus ochrogaster*, one of the few animals thought to engage in genuine monogamous pair-bonds. Many 'monogamous' species, such as migratory birds, only pair up for a single season. Come next summer, they find another avian to shack up with. But the prairie vole 'mates for life'.

Jacked-up genes in this little rodent result in elevated levels of the hormone oxytocin (the same hormone released by orgasm and breastfeeding, popularly known as the 'cuddle chemical')

17. Few animals on the face of the planet are remotely faithful. The prairie vole is an exception – and the clues to its monogamous roots may be found in its genes.

and vasopressin (a multitasking hormone, crucial for constricting blood vessels, helping kidneys to retain water, and also now known to play a key role in social bonding).[87] If its promiscuous relative the montane vole (*M. montanus*) is dosed with the high levels of vasopressin and oxytocin that are found in its monogamous prairie relative, the wanton montane rodents suddenly adopt the settled-down lifestyle of their relatives on the grasslands.

Recent studies of the genetics of these monogamous prairie voles have added a neat neurochemical twist to our understanding of the mechanics of monogamy. The key to their pair-bonds lie not just in their hormones, or even their genes, but in the molecular players – known as 'epigenetic factors' – that control *how* those genes are used. Epigenetics is a new and game-changing field of biology that is rewriting the textbooks on genetics. It turns out that a huge number of molecular actors in the cast of chemical characters can 'turn' a gene 'on' or 'off'. Individual genes code for proteins – the building blocks of your body – and surrounding each gene on its home stretch of DNA are a variety of 'control regions', which control when and why that gene is 'read' and translated into protein. Those control regions can be manipulated chemically (such as by the addition of cumbersome 'methyl' groups that make them difficult to access), which in effect 'silence' a gene. Conversely, there are also chemical factors that enhance the transcription of a gene.

If a genome is a library, and a chromosome is a book, imagine that a gene is a paragraph. An 'epigenetic' factor is like a post-it note flagging up a specific passage – or a piece of gum preventing you from reading the entire page.

Alternatively, if your genome is the cookbook for your body, you can think of epigenetic changes as bits of sticky icing that have glued two pages together, bookmarks here and there, and rough pencil markings scratching out key instructions. Epigenetic changes to your DNA alter how your genes are expressed. And the changes can be passed on through generations. In one of the more unsettling studies, epigenetic changes[88] in holocaust survivors have imprinted on their children a greater susceptibility to stress. Man hands on misery to man in deep molecular ways we are only just beginning to understand.

Molecular detective work[89] indicates that prairie voles pair-bond not simply because their genetics predispose them to monogamy, but because the act of mating itself actually triggers changes in the ways their genes are used. Analysis of the brains of animals who had just mated for the first time revealed that their brains bore higher numbers of receptors for the hormones oxytocin and vasopressin. It is these receptors that actually allow the hormones to work their magic. Without these keyholes, the hormonal keys would be useless. Drugs that enhanced the use of those genes produced the same changes in the number of keyholes in the brain. Hence, epigenetic changes seem to be the chemical clincher that produces the phenomenon we call 'mating for life'.

'The real question has always been: how can mating affect the preferences of the animal forever? In a matter of hours a prairie vole will go from being intensely social to fiercely aggressive to all animals but its mate,' explains Professor Thomas Insel of Emory University, Georgia, who also studies prairie voles but was not involved in this particular study. 'This is a beautiful example of how a behavioural experience is transduced molecularly into a long-term output in patterns of behaviour.'[90]

Could similar genetic triggers in our own species induce the biological changes that inspire devotion to just one person? Certainly, love can certainly feel like one's brain has been rewired, hacked by a neurochemical virus.

## SPERM BOMBS

But while some biologists assert that monogamous behaviour defines our species, others argue precisely the opposite: that the natural nature of human mating is promiscuous.

'Men are just an evolutionary experiment being conducted on by females. Females use men to compete with other females,' says Professor Gordon Gallup, an evolutionary psychologist at the State University of New York in Albany.[91]

Women who engage in 'extra pair copulations' – essentially,

trysts with men who are not their material provider – have the potential to create children on the sly with a larger number of men, diversifying the genetic makeup of their kids.

'We ultimately are inherently out of touch with the evolutionary mechanisms that have evolved to drive our behaviour – in many respects, the cortex is just along for the ride,' he says. 'Many of the really significant problems that humans confronted during evolutionary history have been solved by behaviours that operate below our level of awareness.'

We would thus be driven to be unfaithful not just for the excitement of new bag, but the genetic benefits of diversifying the apples in our basket. If a woman has three children with only one man, he says, she will have sampled 87.5 per cent of his genes. 'Any further children will thus become increasingly redundant copies,' he says. 'Therefore, having children increases the range of potential solutions to future unanticipated environmental problems.'

Gallup argues that female unfaithfulness has been a key driver in the evolution of both human sexual behaviour and our anatomy. Many species of animal have penises that bear spines, spoons, and other tools to scrape out the sperm of rival males before implanting their own. Gallup believes the human penis is also a sperm displacement tool: the coronal ridge (the rim of the bell, if you will) evolved as a means to scour a woman's vagina of rival male sperm, pulling it back out of the vagina and spilling it beyond the reach of her eggs. In his widely cited paper,[92] he records how he and his undergraduates used a range of artificial phalluses (read: dildos) and, using artificial vaginas (read: honeypots), filled the vaginas with artificial semen (fashioned from cornstarch and water), and measured the volume displaced by artificial coitus.

How did they measure the volume of semen displaced? With

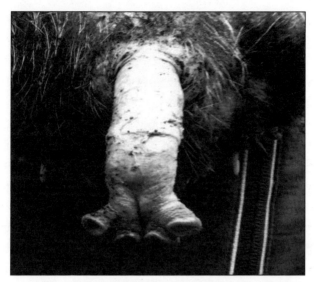

18. The two-pronged hemipenis of an echidna. How many men reading this book would love to spend one day enjoying the possession of a two-headed penis?

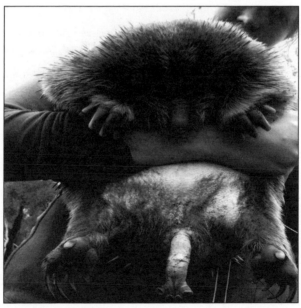

19. An echidna with his aroused member on display.

an equation that would have rendered mathematics lessons far less dull for many a secondary school student:

$$\frac{\text{(weight of vagina with semen – weight of vagina following insertion and removal of phallus)}}{\text{(weight of vagina with semen – weight of empty vagina)}} \times 100$$

'The human penis has evolved to effectively displace rival sperm,' Gallup asserts.

Other biologists have suggested that as a form of sperm-based counterstrike, it could be possible for remnant sperm from the last mating to take up residence in the inviting recesses of a man's foreskin, only to be implanted into the next woman he dallies with. In other words, a man about town could unwittingly piggyback semen from one vagina to the next.

A huge variety of animals compete for access to eggs not via internal claws, horns or teeth, but via internal 'sperm competition', leaving the job of waging sexual warfare to their tiny wriggly soldiers.

A multitude of animals ejaculate sperm that hook together and by joining forces swim at higher speeds towards the ovarian goalpost, from deer mice[93] to koalas and a variety of insects. There is strength in numbers: short-beaked echidna spurt sperm that bundle together in packs of 100 at a time (ejaculated from a penis that bears two heads, dubbed a 'hemipenis').[94]

Many animals, including spiders, reptiles and chimpanzees, simply apply an architectural rather than militaristic solution: they fill females with a 'sperm plug', preventing the next suitor's sperm from gaining entrance.

Do humans unwittingly engage in spermicidal warfare? A quirky analysis in the 1970s[95] of the qualities of subsequent spurts that men produce (anywhere from three to nine per orgasm, didn't you know) suggested that the last spurts (but not the first) contain spermicides that wage war on newcomers. An even stranger 1995 report[96] from Australia declares that men who view pornographic films of two men having sex with one woman produce higher-quality sperm than men who view images of women alone, implying that testicles are primed to produce feisty sperm when it appears the competition is fierce. Make of these findings what you will.

## 'A COLD, DARK SEA'

Others go even further, arguing that monogamous pair-bonds are merely the cultural hangover left by centuries of oppression by church and state – the church wishing to suppress our desires, the state hoping to retain established systems of property inheritance securely in place.

Best-selling book *Sex at Dawn* (2010)[97] is the best-known work to outline the argument in favour of the 'polyamorist' view of sexuality: that humans are built for multiple partnerships. Authors Christopher Ryan and Cacilda Jetha are evolutionary psychologists – as we have noted, a discipline prone to bias and statistical skewing.

In any case, you can make up your own mind about Ryan and Jetha's main idea: 'With and without love, a casual sexuality was the norm for our prehistoric ancestors.'

The nub of their argument: 'The campaign to obscure the true nature of our species' sexuality leaves half our marriages collapsing under an unstoppable tide of swirling sexual frustration, libido-killing boredom, impulsive betrayal, dysfunction, confusion, and shame. Serial monogamy stretches before (and behind) many of us like an archipelago of failure: isolated islands of transitory happiness in a cold, dark sea of disappointment.'

Ryan and Jetha make a number of convincing points to support their view. They point out that bonobos, also known as the 'hippy chimpanzees' – just as close to us on the evolutionary tree as chimpanzees – engage in constant sexual trysts with one another, across all sexes, ages, and even within familial relationships. They employ constant and varied forms of fornication to neutralise conflict and solidify social bonds (making love, not war, like their more bellicose chimp cousins). Bonobos, like ourselves, engage in face-to-face mating (among every other form of copulation imaginable).

In many cultures multiple sexual partnerships are the norm, such as the Aché of Paraguay, who believe in 'partible paternity' (the notion that more than one man has contributed biologically to the birth of a child), or the Marind-anim people of Melanesia who believe that semen is essential to the fertility of a woman. They stipulate that all the men in a village lie with the bride on her wedding night. Ryan and Jetha also describe in detail the Mosuo of China, a marginalised ethnic minority in which property passes through women, who are encouraged to have multiple partnerships.

Ryan and Jetha argue that, among primates, 'body size dimorphism' – or a noticeable difference in size between male and female – is linked to male-dominated harem structures, such as in gorillas, in which one high-ranking male holds court over a group of females while subordinate males go wanting. When males and females are close in body size, egalitarian promiscuity should predominate. They even make a nice tidy chart showing how big our penises are and how close men and women are in body size compared to other apes, and therefore argue that this implies humans are born to be promiscuous.

'Adult male humans have the longest, thickest, and most

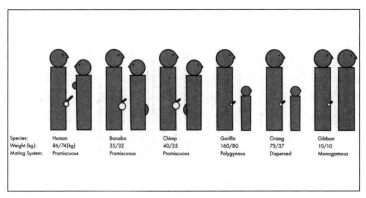

| Species: | Human | Bonobo | Chimp | Gorilla | Orang | Gibbon |
|----------|-------|--------|-------|---------|-------|--------|
| Weight (kg): | 86/74(kg) | 35/32 | 40/35 | 160/80 | 75/37 | 10/10 |
| Mating System: | Promiscuous | Promiscuous | Promiscuous | Polygynous | Dispersed | Monogamous |

20. Penis size, according to the authors of *Sex at Dawn*, correlates not only with the propensity of a primate species to be promiscuous, but also whether they have sex face-to-face or from the rear.

flexible penises of any living primate. So there. *Homo sapiens*: the great ape with the great penis!' They even suggest recent studies have shown the size of male testicles to be declining in size at the moment, and claim 'social monogamy itself may be shrinking men's balls'.

Ryan and Jetha made quite a splash.

Anything popular and controversial will therefore court discontent. There are many counterarguments to their position, like that of Lesley Hall:

'So what?' she sniffs. 'So what if polyamorous multiple relationships were normal two million years ago – that doesn't mean that's the way we live today. I find any argument that says "nothing has changed since the Savannah" to be very problematic.'

*Sex at Dawn* is certainly not the last word on the subject. If the history of sex research or the popularisation of science has taught us anything, it is that debates concerning the 'natural' state of human reproduction will continue to rage unabated, every camp gathering up the data that justifies their point of view and peddling their notions under the banner of 'science'. Studies positing grandiose claims concerning the essence of the human condition will continue to be published, academics will undermine and belittle each other's beliefs in far from civil language, and meanwhile our own species will continue to do as it does: getting married, sleeping around, having babies, breaking up, and falling in love.

Monogamy is celebrated by some as a cornerstone of our biological nature and derided by others who deem it nothing more than a cultural construct – both with equal fervour. 'Polyamorist' pontificators today preach the 'naturalness' of multiple partnerships with the same intolerant evangelical zeal as Christian crusaders who claim lifelong monogamy and abstention before marriage the basis for a happy, healthy, normal human life.

If anything, the study of the human species has taught us that variation is the norm – and this applies to nothing more so than to sex. Some like boys, others like girls, many like both. Whips tickle the fancy of some, while others are set alight by an enema. Missionary 'vanilla' sex, despite being maligned as boring, remains a firm favourite. Some of us – without a doubt – are not built for monogamy, and will never be happy in a lifetime with one person. But to assert that lifelong love does not exist is not only folly – it's downright unscientific. If anything, two centuries of sex science have taught us that diversity is the norm and exception is the rule. There is no 'one size fits all' state that suits everyone. And there never will be.

## MIND–BODY DANCE

Scientists have revealed the unimaginable complexity of our bodies, developed new ways to enjoy sex without pregnancy, traced the invisible ways scent makes someone irresistible, and continues to stimulate and inspire our understanding of evolution's gift.

But an obvious question could be asked: do we really need science to make sex better? Surely the ancients were having a pretty good time long before fMRI scanners hit the scene.

As with most things in science, the main point is not what we have learned – it's what we have to learn.

Professor Meredith Chivers[98] of Queen's University in Canada thinks the science of sex – and, most especially, sex in the mind – is a worthy object of scrutiny.

Chivers' studies patterns of arousal. Simply put, she shows a variety of images to people as scientific instruments measure their degree of excitement. Men will have a glass chamber, dubbed a 'plethysmograph', strapped over their limp penis, and then shown a variety of images, from pornographic films to landscape paintings. The amount of air displaced by their inflating member is catalogued by computers.

Women will don a similar device, essentially a glass dildo with a camera and a light inside. The amount of vaginal lubrication they produce can be measured with optical analysis: bounce light off the surface of the glass, and depending on how much fluid has coated the outer surface, the reflected rays will differ.

Her results never fail to make a splash with the media: while straight men experience erections at the sight of naked women or heterosexual pornography, women experience varying degrees of arousal to a wide variety of images, from straight porn to lesbian porn to wildlife documentary footage of rutting animals. Though straight women do not claim to be aroused by images of naked women, the data seems to indicate otherwise.

'Standard reports on my work tend to paint women as being liars, or completely out of touch with what turns them on, that they are all bisexual, or want to have sex with animals. Or that they are genuinely turned on by sexual violence. Oh, "women are so wacky!" seems to be their take-home message. Those are all very remedial understandings of how our bodies and our identities are interrelated.'

Arousal in the body does not translate to arousal in the mind – and that is what has kept her interested. 'One of the most compelling things about sex is that so much comes from the parts of our nervous system we have no control over,' she says.

Through sex we can explore the complex dance between mind and body, and discover strange new truths about our very strange species. In other words, it is in the animalistic part of ourselves that we discover some of the most startling ideas about what it means to be human.

# DRUGS

# BOTANICAL BATTLES

For an exploration of the history of drugs and science, we ought to start somewhere colourful. We are downright spoiled for choice. The history of the human love affair ranges from a president of the Royal Society's addiction to laughing gas, the world's first acid trip commencing in the mind of a biochemist riding a bicycle on a mountaintop in Switzerland, the shack a rogue chemist set up in his garden where he churned out ecstasy and hundreds of new psychoactive drugs, spiders on speed and monkeys dosed with DMT and locked in dark boxes. Or perhaps, most eccentrically, the hippos that now run rampant in the rainforests of Colombia thanks to infamous drugs baron Pablo Escobar, who imported the aggressive amphibious African beasts for his private zoo.

The history of human drug use is colourful, complex and confusing. Where to begin? Let's start with the simplest question.

What, exactly, is a drug? Definitions vary, and the implications are important.

Some classify a drug as an illegal compound that produces psychological or emotional effects. But all of our favourite legal drugs – tobacco, alcohol and coffee – have at one time or another been prohibited. Today we remember that alcohol spent much of the 1920s as a banned substance in America (bootlegged booze was in fact the source of much of the fortune of the Kennedys; being Irish Catholics, we can assume abstinence from alcohol would have been unconscionable), but we have largely forgotten that the caffeine contained in the fruiting bodies of the plant *Coffea arabica* was once deemed so detrimental to the moral fibre of human beings that coffee houses were outlawed in sixteenth-century Turkey.

Smoking to absorb the nicotine produced by the leaves of

21. Poster for the propaganda film *Reefer Madness* (1937), which has become a cinematic classic adored by potheads everywhere. You literally have to scrape the resin off the seats after a screening.

*Nicotiana tobaccum* through the lining of our lungs widely garnered public floggings in Renaissance Europe – or worse. Being
caught with a cigarette in your hand would earn you a slit nose
in parts of Russia in the seventeenth century. The legal status
of *Cannabis sativa* is changing so rapidly worldwide there is no
point in trying to keep up to date with the pace of change in the
pages of a book. Yet only half a century ago it was portrayed as
a corrupting poison that would lead to murder, theft and inter-
racial sex (heaven forfend).

Suffice it to say that the legality of a substance has little to do
with whether or not it is a 'drug'. So what designates a chemical
as narcotic?

If you were a plant instead of a politician, you would think of
a drug as this:

'Something I use to make animals do what I want.'

Many drugs evolved in their plant producers as a means to
manipulate the behaviour of predators, mainly by tasting foul
to deter insects – or outright poisoning the pests and removing them from the environment. The caffeine in coffee and tea,
nicotine in tobacco, and crystalline tropanes in the coca plant all
developed in their plant hosts as a means to repel insect invaders.
We have inherited this neural hardware from hundreds of millions of years of evolution, so botanical chemicals can tickle our
nervous systems by hijacking the same basic circuitry.

So if these astringent chemicals are designed to repel, why
are we drawn to them? In doses designed for most predators of
plants, these chemicals are poisonous: nicotine, for example, is
lethal to nearly all animals except primates. Birds are highly sensitive to it, and death by tobacco butt is a frequent cause of avian
expiration in modern cities. We are just less overwhelmed by the
effects. As Paracelsus (Philippus Aureolus Theophrastus Bom-
bastus von Hohenheim, to be precise, 1493–1541) the founder

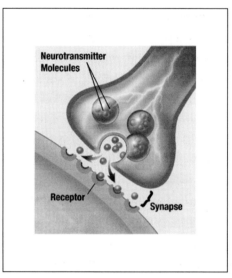

22. Counter-intuitively, nerve cells do not neatly fit together like electrical wires but are separated by spaces called synapses. Chemical messages must traverse this space before information can be relayed via electrical signals.

of toxicology put it half a millennium ago: the dose makes the poison. A bit of something bad can feel mighty good.

Paracelsus made his astute observation without our modern understanding of the mechanics of the nervous system. Had he known then what we do now – that information travels via electrical sparks that zip down the highways of the nervous system, triggering the release of globular chemical messengers that neatly fit into docking bays at the gates of the next highway, in turn triggering the firing of yet more electrical zaps – he would be mighty impressed.

All the nerves in our bodies (if found in the brain, termed 'neurons') are like tiny wires. Akin to electrical cables wrapped in rubber, they are cloaked in an insulating membrane of fat that allows electrical signals to twitter down them at maximum speed.

The velocity is impressive: it allows the sight of this *word* to be transmitted to the patch of your brain that decodes signals from the eyes almost instantaneously. Where nerves connect to each other, a space exists between the nerve cells, dubbed a 'synaptic gap'.

Instead of transmitting this electricity through a continuous string of wires (which might make more sense), nerves have to pass signals to each other by means of globs of chemicals. These have to travel through the synaptic space between two cells like cargo chucked between two ships.

The cargo in these packages are known as 'neurotransmitters': they comprise the chemical language of the nervous system. The docking bays on cells that receive them are called 'receptors'. It is these neurotransmitter messengers that allow electrical signals to be converted into chemical signals and then back into electrical signals.

You've heard the names of the most famous neurotransmitters before: serotonin, dopamine, endorphins and adrenaline are all well known. But there are many thousands more floating around your body that play important roles. The complexity of the circuitry and the intricacy of the chemical lexicon is staggering.

Many drugs work by mimicking our body's native neurotransmitters. For example, the opiate painkiller morphine produced by the poppy *Papaver somniferum* is a better fit for our body's 'μ-opioid' endorphin receptors than our own endorphins are. LSD is a more ardent match for our serotonin receptors than serotonin itself.[1] Ditto for psilocybin,[2] the active component in magic mushrooms. Serotonin receptors latch on to it with a fervent passion, leaving native serotonin without a receptor to call home. It's an impressive feat of biochemical insurrection.

We in fact identified many of our most powerful neurotransmitters and elucidated their structure in the very first place thanks to our relationship with narcotic drugs.

The thought process of scientists in the nineteenth and twentieth centuries went something like this: how can foreign chemicals produce effects in our body? What could be the mechanism at play? Chemists identified the structure of most of our favourite botanical drugs in the nineteenth century – including opium, cocaine, caffeine and nicotine – but how exactly these substances hacked our neural hardware remained mysterious.

Scientists achieved the very first discovery of a neurotransmitter receptor thanks to our predisposition to smoke: biologists at the beginning of the twentieth century were puzzled by the fact that nicotine, derived from a plant, can affect the way an animal feels. Nicotine had first been isolated in 1843, and its chemical structure revealed in 1893. But how this molecule actually produced effects in our bodies was unknown. In 1905 Cambridge professor John Langley proposed that animal tissues must contain 'a substance that combines with nicotine ... [and] receives the stimulus and transmits it.' He called this hypothetical keyhole a 'receptive substance'.[3]

The idea proved insightful and inspirational. It took many decades of biochemical detective work by scientists across disciplines and around the world, but the receptor was finally located by French scientists in the 1970s, who cleverly combined snake toxins with the reactive tissues of an electric eel. The receptor normally binds to our native neurotransmitter acetylcholine, but nicotine can masquerade as an indigenous chemical messenger. The receptor was respectfully named after our relationship with tobacco, and christened the 'nicotinic acetylcholine receptor'. Its sibling receptor, the 'muscarinic acetylcholine receptor', which also normally binds to acetylcholine, is named after fly agaric mushrooms, *Amanita muscaria*. Psychedelic chemicals in this horrifically hallucinogenic fungus bind to the receptors for acetylcholine in a similar act of biological break and entry.

23. The ancient 'hagfish' deserves its jacket description.
The bottom-feeding scavenger not only smells remarkably
foul, it exudes putrid slime from every pore.

According to pharmacologist Professor Richard Miller of
Northwestern University in Chicago, the discovery of the nicotinic acetylcholine receptor was the 'initial breakthrough in our
understanding of the molecular properties of receptors.'[4] Tobacco,
the cause of death for untold millions, led to a watershed moment
in our understanding of the brain.

Similarly, neuroscientists identified our native neurotransmitter painkillers, the endorphins, thanks to the search for the
cellular cradle into which invading opiate molecules fit. Though
opium was a favoured tonic of the ancients, and the chemical
structure of morphine mapped in 1805, it was not until 1973 that
the receptor for opium was located by American neuroscientist
Sol Snyder.[5] He demonstrated that the keyholes for opiates exist
not only in human tissue but also in ancient animals such as the
hagfish, a phenomenally ugly and slimy eel-like marine creature.

Lacking both a spine and a jaw, the primeval hagfish – known as a 'living fossil' – evolutionarily predates fish, frogs, birds, and, indeed, all vertebrates. It is a formidable and rather foul relic.

That receptors for morphine – a comparatively modern molecule, produced by terrestrial plants – could exist in such an ancient animal inspired a ground-breaking idea. Could opiates operate by hijacking the receptors for naturally occurring heroin-like molecules that animals themselves produce? Do our bodies create their own painkillers?

In 1975 they were found: we know them today as endorphins.[6] We may never have found them were it not for our fondness for junk – or the poppy's unrivalled capacity to medicate the human condition. Try as we might, all the laboratory manpower combined with pharmaceutical finance has never produced a painkiller as effective as morphine. Plants are still the victors in the battle between the leaf and the brain.

Moreover, not only did morphine lead us to discover our inborn endorphins, this landmark discovery paved the way for Snyder to also discover the receptors for dopamine – the 'pleasure chemical', one of the world's most famous neurotransmitters – and GABA, one of the most ubiquitous and important neurotransmitters.

In this same vein, the active ingredient in marijuana, the endocannabinoids,[7] is nothing more than a chemical mimic of our body's own painkillers, which are thusly named thanks to our pre-existing knowledge of and relationship with marijuana. The chemical investigation followed the same framework: One, identify the chemical structure of a narcotic drug. Two, locate the receptor that embraces the illicit invader. Three, find the 'endogenous' chemical – native to our own body – that latches on to the same receptor. Voilà: chemists identified delta-9-THC as the active treat in weed in 1964, located the receptor in 1990, and tracked down our own 'endocannabinoids' shortly thereafter. We understood the chemistry of marijuana before we discovered our body's homegrown anaesthetics.

From the perspective of *C. sativa*, evolving the capacity to produce delta-9-tetrahydrocannabinol, or THC, was the best thing that ever happened to it. Compare the global distribution, population and living conditions of *C. sativa* and its cousin *C. indica* with their non-THC-containing relatives. Behold a stark contrast between scruffy scrubs confined to a few remote corners of central Asia and a widely worshipped weed whose seed has spread to every corner of the earth.* Animal emissaries may have unwittingly aided in transporting *C. sativa* and *C. indica* seeds on their hooves, wings and fur, but no animal has done more to spread the progeny of this plant than *Homo sapiens*. So who exactly is domesticating whom?

Evolution has more tricks up its sleeve designed to addle our minds and manipulate our behaviour. Again, plants are the best lockpickers of your genome.

Caffeine produced by coffee, tea and chocolate played a key role in helping scientists unravel the structure of the chemical coinage we use to transmit energy: adenosine triphosphate, or ATP. Known as the 'molecular unit of currency', ATP is a nimble chemical package that all living things use to store and transport energy.[8]

This handy storage battery of a molecule is constantly on the move, swirling round our cell's metabolic circuits. It is continually broken down to release energy to fuel chemical reactions and then recreated to store the energy produced by other molecular transformations in the assembly lines of our microscopic chemical factories.

---

* Except for Antarctica – but undoubtedly scientists have transported the leaves there for personal use.

A mobile and multifunctional molecule, ATP is biological bullion. Universally employed by all forms of life on earth, it is the gold standard of evolution. It never depreciates in value. ATP was officially discovered in 1929 – but the structure of its chemical saboteur, caffeine, was first revealed in 1881, garnering Emil Fischer the Nobel Prize in 1902.

Again, we identified the botanical trickster before becoming familiar with our own friendly biochemical natives.

Intriguingly, caffeine – remember, embedded in the leaves of tea and coffee plants as a means to deter pests – could actually be used by some plants as an alluring attractant, not a revulsive repellent.[9] Dr Geraldine Wright of Newcastle University trained bees to associate a scent with a sugary reward. She found that lacing sugary water with caffeine enhanced the memories of her bees: those fed from caffeine-tainted dispensers were twice as likely to remember the scent (measured by whether or not they stuck out their tongues in anticipation of a sweet reward) than those fed without caffeine. The neurons responsible for memory formation seem to be behind this mechanism: caffeine blocks the receptors for the neurotransmitter adenosine.[10]

This hints at an interesting proposition: can an insect become addicted to a drug? Because caffeine is found in the nectar of more than 100 plant species, Dr Wright thinks it's plausible that plants are dosing their pollinators to coax them into revisiting their flowers and thus spreading their seeds. There are intriguing implications for how caffeine may affect the ability to form new memories in humans, as the scientific evidence for whether or not caffeine is a memory aid is mixed.[11]

At the core of the debate lies this single question: Is caffeine a drug?

You probably know that Coca-Cola originally contained hefty doses of cocaine before the company was forced to remove it.

In fact, the 'secret' recipe today is still made with decocainised coca leaf (which might explain why it tastes better than Pepsi).[12] But few of us remember that long before the company became embroiled in scandals in the developing world over public water supplies, the all-American drinks company endured a lengthy court battle to prove that caffeine is not a dangerous drug.[13]

The battle between plants and insects continues to produce new drugs, but now straight from the laboratory rather than extracted from the leaf. We have unwittingly produced new narcotics as a by-product of our indirect efforts to aid plants in their war with insects. The synthetic drug 4-methylmethcathinone (4-MMC), also known as mephedrone, 'miaow miaow', and M-CAT, was first synthesised in 1929[14] but mostly forgotten for 70 years. Only when Israeli scientists investigated the potential to use it as an insecticide in the early 2000s were its neurotoxic and (in the opinion of some) pleasurable properties discovered when their fingers accidentally became tainted with the chemical's residue. Hence the drug's other nickname: 'plant food'. M-CAT, like Viagra, demonstrates that new and powerful drugs are often discovered by accident rather than deliberate chemical intent.

## ANIMAL NATURE

In the running battle between the nervous systems of animals and the chemical production factories of plants, animals have learned to convert aversions into affections. Reports of wildlife consuming psychotropic, hallucinogenic, stimulating or sedating plants for purposes that seem designed for pleasure alone are rife in the scientific literature.

24. The red toadstool, *Amanita muscaria*, or 'fly agaric'. Unlike other 'magic' mushrooms, this fungus creates hallucinations not through psilocybin but a cocktail of chemicals.

Botanists and zoologists have extensively documented animals deliberately intoxicating themselves. Ducklings too busy feeding on narcotic plants to respond to their mother's calls.[15] Hawkmoths gorging on the nectar of *Datura* flowers.[16] Pumas gnawing at the bark of the *Cinchona* tree are especially noteworthy because indigenous populations in Peru who noticed the behaviour mimicked it themselves.[17] Centuries later, quinine was isolated from the bark in 1820 and deployed in the battle with the parasitic plasmodium viruses responsible for malaria.[18]

Housecats are notorious fiends for the plant *Nepeta cataria*, also known as catnip. Interestingly, wild tomcats are less fond of the herb, and pumas, lions, and other wild species belonging to the group of animals known as *Felis* (cats) are positively averse to it. A captive tiger once 'leaped several feet into the air, urinated, and ran head-first into the wall of his cage upon simply sniffing the leaves'.[19]

The tales are endless. Wild bighorn sheep scamper along treacherous mountain ledges in pursuit of psychotropic lichen.[20] Reindeer are drawn like clockwork to *Amanita muscaria* – also known as fly agaric mushrooms, or red toadstools – seasonally gorging on the fungi and wandering erratically from their migratory paths.[21]

Their ardent attraction to the metabolites of the white-spotted fungus is so strong they will smother themselves in urine left by the people of the Arctic circle who have themselves snacked on fly agarics – even willing to do battle over access to the urine-stained snow. And of course there is the *Sclerocarya birrea* tree, known more popularly by the butterscotchy liqueur Amarula created from it. The fermented fruit draws African land mammals – most famously elephants (hence the bottle's label) – to it in aggressive, rowdy hordes.

Animals in the wild habitually consume intoxicating, hallucinogenic and sedating chemicals. But it is at the intersection between animals, humans, and our own drugs that the story starts to become more interesting. And more questionable.

Take the popular alkaloid nicotine (alkaloids being a broad class of chemicals that includes cocaine, caffeine, morphine, and many other of our favourite drugs). The product of the leaves of *N. tobaccum* is lethal, as noted, to almost all animals. Except primates. The tiniest doses of the chemical will kill insects, frogs and birds. Yet monkeys and apes are somehow impervious to nicotine's lethal properties.

You've probably seen old film footage from the 1920s of circus chimps smoking cigars (perhaps while roller skating).

Chimps have not only been coaxed into smoking – they have become genuinely addicted. Zookeepers worldwide have struggled to get captive animals to kick the habit, most famously Charlie[22] of the Mangaung Zoo, in Bloemfontein, South Africa, who first became hooked when visitors tossed him lit cigarettes.

25. Tobacco companies notoriously hijacked the ideals of Suffragette movements with ad campaigns that framed the right to smoke as a feminist cause.

26. *Monkey see monkey do*. Chimps – indeed all other apes and monkeys – are just as susceptible to the seductive qualities of tobacco as we are.

Enticing captive apes to smoke is an old and clichéd trick. Chimpanzees have been put on show smoking for centuries. The first recorded instance in Europe took place in The Hague in 1635.[23] It was around this time that the dosing of animals took on a scientific, systematic dimension as medical doctors, chemists and other enterprising experimentalists of the burgeoning scientific disciplines began to immerse, inject, feed and cloak animals in a variety of psychoactive chemicals, from alcohol to ether, morphine to mescaline.

For better or for worse, the use of animals to investigate the properties of drugs resulted in a number of medical milestones. The first needle (which eventually evolved into the heroin-friendly hypodermic form) was produced by Sir Christopher Wren (1632–1723), who paired goose quills with animal bladders to inject dogs with opium in 1656. Another breakthrough came with the French medic Pierre-Alexandre Charvet (1799–1879), who in 1826 published *Action comparée de l'opium et de ses principes constituans sur l'économie animale*,[24] a detailed account of his work administering opium to animals, including paramecia, crayfish, snails, fish, salamanders, frogs, birds, rabbits, dogs, cats – and himself. This is regarded by many as the first book in the field we now know as 'experimental pharmacology'.

LSD has frequently been given to animals. Guppies dosed with acid swim into the walls of their tanks.[25] Siamese fighting fish display their fighting stance to unoccupied water.[26] Worms work their way upwards through soil.[27] Snails fall from the sides of trees.[28]

Perhaps the most famous experiments on animals with LSD involved spiders, who were dosed with a variety of narcotics[29] in experiments funded by space agency NASA.[30]

Why arachnids? In addition to being cheap and easily sourced, the influence of a drug on the arrangement of a spider's web seemed to provide a rough indicator of a chemical's toxicity.

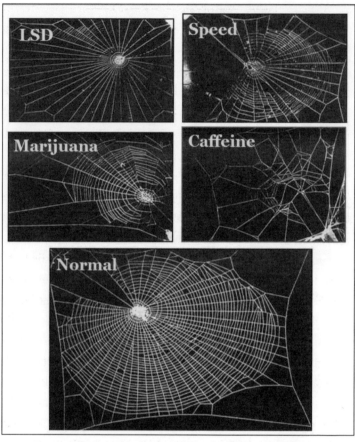

27. Even spiders can get high. Studies funded by NASA tested a variety of drugs, from quotidian caffeine to other worldly mescaline, by examining the impacts of each narcotic on web geometry.

The geometric array is illuminating: webs made on caffeine are erratic and nonfunctional. Chloral hydrate, also known as the sleeping aid Hydrate, produces a sparse arrangement of threads. Marijuana results in a fairly competent web, but it appears as if the spider abandoned its task halfway through. The LSD web resembles a normal one proportionally, but it is definitively broader. Make of this what you will.

Aside from sheer spectacle, what is the utility in experimenting on animals with drugs? Non-human models cannot make for a perfect replica of the brain of *Homo sapiens*, but according to many biologists, they serve as a decent proxy.

Psychiatrist Dr Ronald Siegel of the Department of Psychiatry and Biobehavioral Sciences at the University of California Los Angeles, spent the better part of 30 years studying the impacts of narcotics on animals. He believed that understanding the biological basis of addiction would aid his treatment of patients suffering the ravages of chemical dependencies. Siegel worked daily as a clinician treating addicts: he saw with his own eyes how drugs destroy lives.

His aims were undoubtedly compassionate and his methods undeniably colourful. He trekked to the mountains of the Andes to understand the ancient use of the coca leaf. He darted a captive colony of chimpanzees off the coast of California with cocaine both at the low concentrations found in the leaf and with the high concentrations found on the street. Result: coca leaf levels rendered the animals social and cheerful. Miami-style doses led to aggression and disrupted social dynamics. Odd as this may sound, it was not until Siegel's experiments that cocaine was accepted by the scientific establishment and the general public as 'addictive'. Until then, it was merely thought to be habit-forming.

His experiments with the hallucinogen *N,N*-dimethyltryptamine (DMT) on rhesus monkeys are described with such striking detail that, rather than paraphrasing, they are best quoted verbatim:

Darkness, solitude and the silence of night are the most common times for humans to use hallucinogens. All primitive societies prefer these drugs under conditions when there is little else to see or hear in the environment... In dark and

28. Geographically widespread and found worldwide in high population numbers, rhesus monkeys have been used to develop vaccines for rabies and smallpox, drugs for HIV, and – in less orthodox settings – to study the effects of the world's most powerful hallucinogens.

isolated settings, monkeys also find exploring visual stimuli exciting and rewarding. In a classic demonstration of this motivation to explore, rhesus monkeys were confined one at a time to a dimly lighted wire cage that was covered with an opaque box. Because monkeys need visual stimulation as much as we do, I was certain they would accept an enlightening hallucinogen rather than darkness. Initially, each monkey was given the opportunity to live alone in the dark chamber for ten consecutive days and nights. The smoking machine was filled with cigarettes made from ordinary garden lettuce … and the cigarettes were laced with DMT … by Day 8 [Claude] had worked up to smoking almost two whole cigarettes each day … Lucy has been smoking almost two DMT cigarettes each day. She has become extremely proficient at catching whatever she has been chasing: she now brings 'it' to her mouth, chews and lip-smacks with delight.

His conclusion: if deprived of light, stimulation, company and comfort, monkeys will smoke an overwhelmingly strong psychedelic. He says this has profound implications for our own species. 'Under the right conditions [DMT] was as useful to a monkey as it is to a human. We share the same motivation to light up our lives with chemical glimpses of another world.' One is tempted to suggest that had Dr Siegel ever tried hallucinogens himself (he maintains that he would never, *never*, take any of these substances), he wouldn't have locked a monkey in a dark box to reach this deduction.

Among his peers, opinions are mixed.

'He's not exactly very popular,' says Rick Doblin, PhD, founder of the Multidisciplinary Association for Psychedelic Studies (MAPS), who has endeavoured to demonstrate the potential to treat human ailments with psychotropic drugs since the 1970s. 'He did tremendously important work, but not everyone is a fan of his methods.'

'I genuinely think of him as a hero,' says medical doctor Dr George Koob, Chairman of the Committee on the Neurobiology of Addictive Disorders at the Scripps Institute in California, who has spent his career trying to understand what predisposes some but not others towards addiction. 'He really was the first person to chart the case histories of cocaine addiction when nobody thought of it as a "addictive drug".'

# 'THE IDEAL DREAMS OF WILD DESIRE'

Rewind from the 1980s to the 1800s and the concoction of the first lab-born intoxicant – and one of the first scientific records of what we would today deem 'addiction'.

The discovery of nitrous oxide ($N_2O$) marks a crucial point in the history of the scientific understanding of narcotics. Records of $N_2O$ experiments constituted the first bona fide systematic investigation of a drug. It was also the first psychotropic compound to reach the human brain that was concocted in the laboratory. Curiously, it is common on planet earth, pervasive in the atmosphere (and a potent greenhouse gas, roughly 300 times more powerful than carbon dioxide in its climate changing capacity[31]). But it was not until chemists figured out how to purify it with glassware and capture it within silken bags that it became available to the receptors of the brain.

Sir Humphry Davy's explorations of $N_2O$ ultimately led to one of modern medicine's greatest achievements: pain-free surgery. His records of the effects of $N_2O$ constitute the first methodical scientific investigation of a drug's properties on human subjects. (Mostly, himself.)

Davy, though he popularised the gas, was not the first chemist to isolate pure samples of $N_2O$. That credit goes to Joseph Priestley (1733–1804), who produced nitrous oxide in 1772 and termed it 'phlogisticated nitrous air'. Alchemists and protochemists of the Renaissance believed in the existence of a fiery form of energy called 'phlogiston', thinking it bestowed on explosive materials their combustible nature.*

---

* The 'phlogiston theory' today is obsolete, but its widespread acceptance at the time illuminates an important facet of scientific progress: accepted truths, no matter how popular, constantly shift.

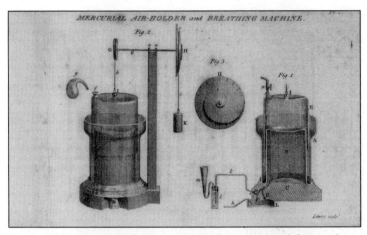

29. A 'mercurial air-holder and breathing machine' designed by Humphry Davy in 1800 featuring a brass axis, glass tubes and cylinders and a moveable receiver. Then – as now – kids created complicated and sophisticated contraptions for the sole purpose of getting out of their heads.

Priestley explored and described the gas in impressive detail in 1775, then passed the chemical baton to Thomas Beddoes (1760–1808) and James Watt (1736–1819). Together they proposed in 1794 that the strange gas could have medical applications in *Considerations on the Medical Use and on the Production of Factitious Airs (1794).*[32] 'Factitious' today is employed to denigrate a concept as false, but Beddoes and Priestley used the term to indicate that the chemical was synthetically produced. Beddoes – after being expelled from Oxford for his atheist and Republican views – established a centre for the betterment of public health outside Bristol for the exploration of gases, drugs and diets.* It was here that he asked young employee Davy to systematically gather data on the gas and explore potential medical applications.

---

* Many may smile wryly to themselves, noting that Bristol was then and still is a hotbed of psychotropic chemical consumption.

Davy, for his part, was perfectly suited to the job: in the burgeoning field of chemistry he saw the potential to understand the 'laws of organic existence'. What matter is made of, and what distinguishes living from non-living.

He took on his task with gusto. Like teens with multichamber pipes, water bongs, and gaseous vaporising devices, Davy spent late nights in the laboratory concocting a stylish range of apparatuses in his quest to consume pure volumes of gas. Metal mouthpieces, silken face masks, tubes capped with corks to be inserted directly into nostrils.

Like most of his peers (yet in stark contrast to today's pharmaceutical magicians), Davy tried everything on himself first. He suffered a few unfortunate accidents with less benign gases, including carbon monoxide (CO). The incident nearly killed him. He decided on $N_2O$ for his devoted study. It is notable that he would always experiment with such dangerous substances on himself first before testing anything on another being – even animals. Only after he was sure a gas was safe would he enlist human volunteers in his chemical forays.

His first adventures with $N_2O$ proved not unpleasant:

A highly pleasurable thrilling, particularly in the chest and extremities ... objects around me became dazzling, and my hearing more acute ... this gas raised my pulse upwards of twenty strokes, made me dance about the laboratory as a madman, and has kept my spirits in a glow ever since.[33]

He spent ever more time in the lab inhaling increasingly large volumes of the gas. Now, most people's descriptions of their drug experiences tend to be unbearably boring. But Davy's recollection – hilarious, honest, and almost entirely lacking in self-awareness – bears repeating:

By degrees as the pleasurable sensations increased, I lost all connection with external things; trains of vivid visible images rapidly passed through my mind and were connected with words in such a manner, as to produce perceptions perfectly novel. I existed in a world of newly connected and newly modified ideas. [*In this we can see echoes from the past of the words of today's psychedelic enthusiasts: everything which once seemed inconsequential suddenly becomes imbued with meaning.*] I theorised that I had made discoveries. When I was awakened from this semi-delirious trance … indignation and pride were my first feelings … and with the most intense belief and prophetic manner, I exclaimed: Nothing exists, but Thoughts! The Universe is composed of impressions, ideas, pleasures and pains!

As his love affair with the gas progressed, he continually amped up the tech. Most impressively he built a portable box – 'like a sedan chair' – in which he could squat while $N_2O$ was pumped into it. He once managed to inhale 80 quarts of the gas – a volume that would put today's most committed users to shame.

While Davy thought something in the gas imbued it with a healing force, other chemists thought quite the opposite. American physician Samuel Latham Mitchell (1764–1831) called it the 'Great Disorganiser', thinking it a lethal counterpart to oxygen.

Like the psychedelic chemists who followed in his footsteps 150 years later, Davy held late-night parties with the cool cats of the day to share his new discovery. He did this not only in the spirit of generosity, but also scientific inquiry, hoping to gather data from new subjects.

Robert Southey* (1774–1843): 'I am sure the air in heaven must be this wonder working gas of delight!'

---

* Poet Laureate for three decades.

Samuel Taylor Coleridge (1772–1834), who himself is credited with coining the word psychosomatic: 'an unmingled pleasure'.

Anonymous: 'I felt like the sound of a harp'.

Oscar Wilde (1854–1900), a hundred years later: 'I felt that I knew everything'. He was under the influence of the gas in the position in which most of us become familiar with $N_2O$: in the dentist's chair, having a tooth extracted.

Notably, Davy did not realise that this would become the most important application for his gas: transforming the experience of putting oneself into the hands of a doctor.

The official honour for the first public use of $N_2O$ as a surgical painkiller is a matter of debate. Some credit American Horace Wells (1815–1848). After removing his own molar without suffering, he was inspired to stage a public extraction to demonstrate the virtues of the gas. But his patient cried out in pain. Wells was publicly disgraced. Discredited and despondent, he demonstrated the utility of anaesthesia in a more convincing fashion: he used a special combination of carbon, chlorine and hydrogen, $CHCl_3$, also known as chloroform, to sedate himself before committing suicide with a razor blade.

Exactly who 'discovered' the power of anaesthesia is a matter of debate, but Davy was not blind to this potential use of the gas: 'As nitrous oxide in its extensive operation seems capable of destroying physical pain, it may probably be used with advantage during surgical operations'.

Though he used the gas to dampen his own wisdom tooth pain, he was interested in a grander use: prodding consciousness and the border between life and death. In $N_2O$, he saw a tool to understand the mind.

Because Davy's public demonstrations focused more on the psychedelic rather than anesthetic applications of nitrous, he found himself the object of public mockery, including a periodical cartoon published in 1802 that depicts the wondrous gas expelling from his backside in a trumpeting cloud.

But we need not pity Davy and his surgical trump by Wells. Davy not only went on to become President of the Royal Society, he also designed the Davy Gas Lamp, brilliant in its simplicity, history-altering for its life-saving capacity.

Until his invention, miners died by the thousands from tunnel collapse when methane gas seams would open and the gas would explode. Methane's lethality is remarkable when compared to its chemical composition: it is one of the simplest molecules imaginable, just four atoms of hydrogen bound to one central heart of carbon: $CH_4$.

Methane is colourless and odourless, yet easily combustible. Miners could not see or smell the gas before it caused uncontrollable eruptions.

Davy's lamp, little more than a metal mesh cage encircling a tiny flame, allowed for controlled continuous burning of methane, preventing explosions. Very clever.

But before he saved the lives of untold numbers of people, the President of the Royal Society stayed up all night for years huffing nitrous oxide and probing the corners of consciousness.

## BEFORE THE BOOMERS

Nothing fun can be kept secret for long. When nitrous oxide escaped into the dim saloons and dinner parties of nineteenth-century Britain, it did not encounter the uptight, emotionally constipated, sober-minded society we envision Victorian England as being. Quite the opposite.

Europe was awash with drugs, having co-opted the narcotics,

stimulants, hallucinogens and intoxicants of the new world – tea, coffee, tobacco, cocaine and raw sugar – with great gusto. The injection of new, quirky and interesting stimulants aligned with developments in chemical analysis in the nineteenth century. This led to the maturation of the scientific study of drugs as we know it: the illumination of molecular structures and the refining of techniques for synthesising pure compounds. Victorian advancements in the science of chemistry, and our ability to create new drugs, are in part thanks to the industrial revolution: the distillation of coal tar in the furnaces of factories produced the sticky by-products naphthalene (isolated in 1820), aniline (1841) and benzene (1845), all of which proved useful to both chemists and druggists. These new carbon-based building blocks allowed for the creation of innovative narcotics.

Most notable: heroin. Opium was an old friend to Europeans – Paracelsus himself developed the tincture laudanum. But the nineteenth century saw the sticky white milk of the poppy *Papaver somniferum* analysed, purified and popularised in unprecedented ways. Opium became increasingly fashionable, and thus predictably led to the first 'drug confessional': *Confessions of an English Opium Eater* by indolent posh boy Thomas de Quincey (1785– 1859), purportedly the most eloquent Latin speaker of his era. Then, as now, smart people are drawn to debilitating chemical stress relief to the point of self-obliteration.

And then, as now, standard-issue drugs proved inadequate for many. Morphine, first devised in 1805, is a refined concentration of the naturally occurring botanics found in opium. And it proved then (and today) irresistible: it remains the strongest painkiller known to man, and thus is understandably unrivalled in its addictive (and destructive) power. Chemists thus sought to devise less lethal chemical analogues. The compound 'Heroin' was concocted by pharmaceutical giant Bayer, who sought to rebrand the chemical diacetylmorphine with heroic, virtuous qualities, divorcing it from the stigma that opiate addicts had acquired. The word 'junkie' in fact is over a century old: the nickname was given to addicts who scrounged scrap metal from junkyards to fuel their habit. In fact, Bayer initially sought to market Heroin as a *non-addictive* alternative to morphine, just as Subutex and methadone are today proffered as preferable chemical counterparts to Heroin. Unfortunately for Bayer (and ourselves), the opposite proved to be true. Heroin is a more efficient means of opiate delivery than morphine. It can more easily enter the brain from the bloodstream than the slightly more molecularly cumbersome morphine. Once in the brain, heroin is instantly converted into delectable morphine. As author Richard Miller puts it: 'Basically, heroin is an intercontinental ballistic missile system for rapidly delivering morphine into the central nervous system.'[34] Whoops.

The late nineteenth century's most popular and exciting drug – exciting because it was new to the Old World – was undoubtedly cocaine. A potent trigger for the release of the pleasurable neurotransmitter dopamine, 'it is simply the nicest chemical gift that the brain can receive,' as Siegel puts it. The scientific lever-point came in 1860 when German chemist Alfred Niemann (1834–1861) isolated benzoylmethylecgonine, the crystalline tropane alkaloid compound that comprises the active component in the coca leaf. By elucidating the chemical, he 'liberated' the compound from the leaf, and pure samples of the powder could be produced.

Niemann's discovery, however, went mostly unnoticed for two decades, until enterprising scientists began to investigate the material. Most notable: psychiatrist Sigmund Freud (1856–1939). He delighted in the compound, and in 1884 published *Über Coca*, expounding upon the 'most gorgeous excitement' it produced. It bears similarity to the writings of 'Dr' Timothy Leary, Harvard

30. *Mrs Winslow's Soothing Syrup*. Today: Gripe water. Then: Opium-infused syrups. One wonders if the tranquillising qualities of opiates rendered the products even more appealing to parents.

psychiatrist turned psychedelic preacher, who tiresomely expounded on the virtues of LSD in the 1960s.

Cocaine's 'most gorgeous' benzoylmethylecgonine alkaloid was infused into drinks, syrups, lozenges, bandages, and even sweets for children. Which merits scrutiny. Did children of the age require assistance in achieving heightened states of hyperactivity? Mrs Winslow's Soothing Syrup, a pediatric treatment for teething children infused with opiates, makes more intuitive sense as a parenting aid.

One of the era's most popular products was Vin Mariani, roughly 10 per cent alcohol and 8 per cent cocaine, beloved by American presidents, Queen Victoria, and two popes, to name just a few.

Coca Cola was famously infused with cocaine and marketed as the 'temperance drink', a sober alternative to alcohol.

The quirkiest vignette from Victorian drug use concerns the

chemical solvent ether, which – if you are a fan of symmetry – has a rather pleasing chemical formula: $CH_3-CH_2-O-CH_2-CH_3$. Though ether compounds were known to Paracelsus and his contemporaries, it was not until the nineteenth century that it became the favoured drug for a select few.

Why? One, like cut rate alcohol today, ether was cheap and easy to consume. All that was needed was a rag and a bottle. Two, like ketamine today, it was short-lived and apparently left no hangover. The effects of intoxication wore off so quickly that etherised bedragglers would be sober by the time they had been dragged to the police station. Three, it served as an alternative intoxicant when high taxes were placed on alcohol.

Draperstown, Northern Ireland, marks the most famous case of a collective ether outbreak. In the 1840s the city's inhabitants en masse fell in love with the solvent's 'triumphal march into nothingness',[35] which included consuming it in liquid form and sharing the drink with children.* Historical accounts testify that visitors could smell the town half a mile away.

From a chemical standpoint, the Victorian age saw a number of milestones, as the expanding field of chemistry consolidated its place among established sciences. Chemists isolated, purified and analyzed numerous psychoactive compounds: caffeine, $C_8H_{10}N_4O_2$, in 1820;† nicotine, $C_{10}H_{14}N_2$, in 1828; benzoylmethylecgonine, the active ingredient in cocaine, $C_{17}H_{21}NO_4$, in 1860; and theobromine, $C_7H_8N_4O_2$, a psychoactive component of chocolate, in 1842. Note that all of these molecules are just variable arrangements of carbon, hydrogen, nitrogen,

---

* One can only imagine the discussions which may have taken place at regional meetings of the educational boards. 'Tut tut, the children of Ether Town cannot raise their GCSE scores...'
† Catalysed by the encouragement of Johann Wolfgang von Goethe (1749–1832), who always expressed a keen interest in chemistry.

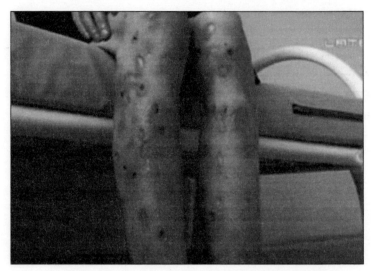

31. Damage caused by 'krokodile', an opiate produced
from the by-products of codeine. On the other side
of the world 'paco', also known as 'cocaine's garbage'
is laying waste to South America's slums.

and sometimes oxygen. Nothing poisonous or radioactive is
required. Nature does not need exotic elements to create intoxicants: all she has to do is assemble commonplace atoms in an
imaginative manner.

Another chemical achievement of the 1800s was the ability to
increase the potency of drugs, made possible by the identification
of their molecular structure by intrepid chemists. The amount of
cocaine in the coca leaves that high-altitude inhabitants of South
America chew for pep is low compared to levels produced in
the drinks, syrups and powders manufactured in Europe in the
nineteenth century. The move to formulate drugs into powders,
solutions and tablets set the trend for what drug purveyors would
replicate continually for the next century and a half in the labs of

pharmaceutical giants and rogue disseminators of banned narcotics alike. Opium begat morphine which begat heroin. Coca
wine was followed by cocaine and crack. Sugar (which we will
return to later) was strengthened and refined when it reached
Europe, where people quickly figured out how to concentrate the
juices of *Saccharum* plants into the high-octane substance we
know and love.

The trend continues: in the 1970s, British chemists discovered
how to concentrate LSD into small beads of solid lysergic acid
distillate, clamped between two plastic squares. 'Microdots' were
easier to disseminate than sheets of paper or cumbersome bottles
of liquid. Chemical trickery can be more caustic and less refined:
dealers to the junkies of Russia have devised ways to produce
heroin-like substances from the residues of opiate pharmaceu-
ticals such as codeine, due to a shortage of the real deal coming
from Afghanistan. The nickname 'krokodile' derives from the
effects the drug produces on human skin, causing it to flake or
fall off in chunks.

One class of drug is remarkably absent from the medicine
chests of nineteenth century Europe: *Psilocybe semilanceata,*
colloquially known as 'liberty caps'. Though the blue-stemmed
fruiting bodies of this fungus are widespread in Europe – and
especially in the soggy UK – all known records of their consumption until the present day were recorded as unpleasant.[36] Despite
modern myths that the roots of British culture were woven by
druids gorging on hallucinogenic mushrooms, accidental inges-
tions of the psychedelic mushrooms (mistaken for edible) were
apparently unpleasant. The best-documented account comes
from 1799, describing the 'J.S.' family, who accidentally brewed
a stew with blue liberty caps plucked from Green Park, mistak-
ing them for edible mushrooms.[37] The youngest of the family,
Edward, was reportedly 'attacked with fits of immoderate

laughter', as described by a physician in the *Medical and Physical Journal* that year.

The first written evidence in western European literature for the use of mushrooms as inebriators derives from Spanish chroniclers in the early 1500s. These reports were greeted with intense suspicion, especially by esteemed botanists. American Dr W. E. Safford claimed that 'sacred mushrooms' did not exist in a speech to the Botanical Society in Washington in 1915.[38]

Subsequent scientists sought out the fungi (once so common in Britain, head shops sold freeze-dried 'mushies' by the bucket) and proved their existence and significance. In 1938 Robert J. Weitlaner and Dr Richard Evans Schultes of Harvard finally located the elusive hallucinogenic mushrooms in the mountains of Mexico. But local Mazatec communities had little desire to share their rituals with white intruders. Westerners had to wait until 1955 for anthropologists Dr Robert Gordon Wasson (1898–1986) and his wife, paediatrician Valentina Pavlovna Guercken (1901–1958), to earn the trust of the Mazatecs. They partook in the experience, and described it in *Life Magazine* in 1957.[39] Three years later Albert Hoffman – the same Swissman of the famed lysergic bicycle ride in the Alps – isolated and quantified the compound psilocybin in the colourfully named journal *Experientia* in 1958.[40]

In a similar fashion, a white European academic botanist brought the powerful drug DMT out of the jungles of South America into western counterculture through a combination of intellectual determination and courageous exploration. The indigenous populations of the Amazon have consumed DMT for thousands of years in a brew called ayahuasca, which combines plant leaves that are naturally high in DMT, typically the visually unimpressive *Psychotria viridis*, with preparations of the *Banisteriopsis caapi* vine, which contain a 'monoamine oxidase inhibitor' (MAOI).

The DMT contained in the leaves of *P. viridis* would be broken down into inactive compounds in our stomachs quickly without an MAOI there to safeguard it. How Amazonians deduced which plants to combine out of the thousands of plant species surrounding them in the rainforest remains one of the brew's most enticing puzzles.

The first westerner to drink the potion is credited as British botanist Richard Spruce (1817–1893). The snuff tray, pestle and tube he used in 1852 are housed in the Economic Botany Collection at Kew Gardens, London. His experiment was profound, but remained an isolated incident for over a century. The brew continued to be exotic, far-flung, and shrouded in mystery. Stories leaking out from the rainforest were understandably alluring: Austrian anthropologist Gerardo Reichel-Dolmatoff (1912–1994) spent three years in the 1960s with the Tukano Indians in the Vaupes area of Colombia because he had heard that shamans swore they became jaguars under the drug's influence.[41] Ayahuasca is now – like cocaine, coffee and tobacco before it – working its way out of the jungles of South America and into the cities of Europe, North America, and beyond, where well-connected cool cats running with the right herd can 'see with the eyes of the panther'.

It is noteworthy that 3,4,5-trimethoxyphenethylamine $(C_{11}H_{17}NO_3)$,* the active component in mescaline, derived from the Mexican peyote cactus *Lophophora williamsii*, was described in scientific papers long before European psilocybin $(C_{12}H_{17}N_2O_4P)$.† Anthropologists had observed the use of the cactus by the indigenous populations of the Americas, and the first use by a white visitor is generally credited as James Mooney

---

* Again, the simple ingredients of carbon, hydrogen, nitrogen and oxygen.
† Mushrooms, distinctly, include phosphorus.

32. November 1969, San Francisco. A march in protest
of the Vietnam War. The military was in fact far more
interested in the potential uses of LSD than most
counter culture cats would have believed.

(1861–1921) of the United States Bureau of Ethnology. In 1891 he
was permitted by the Sioux to join in their ritual.

Mescaline's chemical structure was elucidated in 1897 (60
years before psilocybin) by German chemist Arthur Heffter
(1859–1925). Successful synthesis of the chemical was achieved by
the Germans again in 1919, when the journal *Der Meskalinrausch*
published the chemical structure of the molecule.[42]

In the 1800s scientists were already interested in the potential
use of mescaline as a nerve tonic, heart stimulant, and therapeutic tool. But the 1940s saw the drug's most nefarious deployment:
a probe in the hands of Nazi interrogators. In the never-ending
search for a 'truth serum', both the Third Reich and the Allies
experimented on prisoners of war with hallucinogens. Predictably, the addling psychedelics proved inefficient at extracting

reliable information. Yet the failure of mescaline to prove its worth did not deter the British from trying their hand at psychedelic combat. In 1952 MI6 commissioned a series of trials at the Porton Down military facility near Stonehenge, thinking LSD could be useful as a weapon of warfare.*

Soldiers deployed on routine exercises under the influence of the drug[43] – who mostly just had trouble reading maps and decided to climb trees instead – gave some indication for how the molecule could be useful. Not as a truth serum, but as a mode of destabilising the enemy: a battlefield incapacitant. Hippies would later bear placards stating DROP ACID, NOT BOMBS in protest of American campaigns in Vietnam, but militarists had pondered doing exactly that 15 years earlier.

## OF PARASITES, BICYCLES AND SAINTS

Lysergic acid diethylamide is more commonly known as LSD – or, simply, acid. Although the story of its birth has been told many times, it bears repetition as a scientific milestone.

Many know that LSD, a synthetic molecule, was born in a laboratory in Switzerland and consumed for the first time by a chemist as he rode home on a bicycle. But fewer are aware of some of the more quirky traits of the family of chemicals that lysergic acid belongs to: the 'ergot alkaloids'. All alkaloid drugs are interesting in their own way, but the ergot alkaloids are particularly curious.

One: they are product of a parasite.

Two: they have a saint.

---

* http://www.youtube.com/watch?v=_wAbh3u57hA

Three: they have 'uterotonic' qualities. Ergot alkaloids – including lab-born LSD – can induce contractions in the womb. LSD's wild relatives have been employed to induce birth in delayed and difficult conditions (as well as abortions).

We'll get to the parasite and the saint shortly. Let's start with the lab. Chemist Albert Hoffman was working for Sandoz laboratories in the 1930s and 40s in Zurich, Switzerland, where he had been tasked with investigating the alkaloids of the ergot fungus, known as 'ergotamines', to see if any might prove useful to medicine. Hoffman was not personally familiar at this point with chemical modifiers of the human experience: he had spent his childhood in the Swiss mountains in the most idyllic, healthy and untarnished setting imaginable.

Before turning his attention to LSD, Hoffman was primarily interested in 'plant and animal chemistry', and spent a great number of years working on the gastrointestinal juice of the vineyard snail. 'I accomplished the enzymatic degradation of chitin, the structural material of which the shells, stings and claws of insects, crustaceans and other lower animals are composed,' he writes.

The Swiss chemist found this research so rewarding he devoted his doctoral thesis to the molecular secrets of chitin, the crustacean equivalent of the material that makes up human bones. You come into contact with it every time you crack the claw of a lobster to extract its gooey innards.

Hoffman's employers at Sandoz asked him to turn his chemical attentions to *Claviceps purpurea*, colloquially known as ergot fungus, a black parasite that infests the kernels of rye.

*C. purpurea* produces long, black, phallic botanical squatters that render rye crops toxic. Humans who ingest the infected grain may develop 'ergotism'. This can result in gangrene so severe that hands and feet may atrophy to the point where they fall off. There are historical accounts of peasants who lost all four limbs to ergot poisoning.[44]

This fungus can also produce convulsions and hallucinations, which has led to modern conjecture that humans have unwittingly dosed themselves with LSD-like chemicals for millennia. It has been proposed that the barley contained in the ancient Greek ritualistic potion *kykeon* was tainted with ergot, which would have explained the drink's supposedly enlightening properties. Much has also been made of the possibility that ergot-infested fields were responsible for outbreaks of collective hallucinations throughout European history and therefore that mass LSD trips are responsible for witch-hunts, ghost sightings and other paranormal phenomena. The evidence remains equivocal.

It is, however, certainly true that ergot poisoning was a commonplace affliction in medieval Europe. In AD 944 roughly 20,000 people died in Aquitaine.[45] The last known major outbreak took place in 1926 in southern Russia.[46] The victims of ergot fungus poisoning had their own Christian saviour: St Anthony, the patron saint of ergotism victims. This is also where the alternative name for a lysergic experience, 'St Anthony's Fire', comes from. The gangrenous form of ergotism produces intense fiery burning pains.

The existence of St Anthony reminds us that trips can be bad – hellishly bad – if one is unaware of the cause of the hallucinations.

The uterotonic qualities of LSD are seldom discussed. What is usually recounted is that LSD was initially explored not as a chemical to modify the mind, but as a tool for preventing excessive bleeding.

The first known mention of the medical applications of ergotamines are found in the writings of Adam Lonitzer (also known by his Latinised moniker Lonicerus, 1528–1586). In 1582 he described ways to utilise the fungus as 'an ecbolic' (a medicament

33. The formidable cane toad stands alone among amphibians:
it can tolerate salt water, will readily devour mice and birds,
and will eat dead as well as living matter – even garbage.

to precipitate childbirth). The first description in modern academic scientific literature dates to 1800 in the *Account of the Putvis Paturiens, a Remedy for Quickening Childbirth*,[47] by American John Stearns. The ecbolic properties of the fungus were not advocated by the medical profession at large as the potions pose a lethal danger to the child if prescribed in too large a dose. The use of the fungus was therefore restricted to the prevention of postpartum haemorrhaging and death of the mother by blood loss following birth. This is where the modern simplified description of ergot as a 'vasoconstrictor' derives. Medieval midwives, however, were aware of the molecule's powerful properties, and they applied it in another useful manner: as an abortive.

Investigations into what other uses could be derived from ergot alkaloids began in earnest in 1917, when Arthur Stoll – later Hoffman's superior at Sandoz – isolated the ergotamine (also known as ergometrine) molecule, $C_{19}H_{23}N_3O_2$. Again, same chemical ingredients, nothing more than quotidian carbon, hydrogen, nitrogen and oxygen.

Professor Stoll tasked his laboratory underlings with isolating pure chemical compounds from plant and animal sources because, as Hoffman puts it, 'in the case of medicinal plants whose active principles are unstable, or whose potency is subject to great variation, [this] makes exact dosage difficult'. In other words: identify, synthesise, standardise.

Stoll gave Hoffman the duty of studying foxglove (*Digitalis*), the Mediterranean squill (*Scilla maritima*) and the fungal ergot of rye (*Claviceps purpurea*). It is interesting to note that the glycoside, or sugar-containing, substances of *Digitalis* (digoxin, or $C_{41}H_{64}O_{14}$) bear striking similarities[48] to the toxic compound found in the skin glands of the formidable toad *Bufo marinus* (bufotenine, or $C_{12}H_{16}N_2O$).

Also known as the cane toad, this infamous amphibian has single-handedly decimated populations of countless native animals in Australia, where the toad is an 'alien' or 'invasive' species.

Intrepid Aussies have figured out how to squeeze the poisonous toxin from the toad's glands, allowing it to dry and crystallise, then sprinkling it into a cigarette.

The resemblance between bufotenine and digoxin is an example of a phenomenon biologists term 'parallel evolution': a characteristic found in two species on distant branches of the tree of life. For example, both bats and dolphins use echolocation to navigate and hunt: they emit high-pitched frequencies and use the reflected patterns of sound to map their surroundings and snare their prey. It works just like naval sonar, and evolved in the two groups of mammals independently.

Similarly, the ergot alkaloids found in the fungus *C. purpurea* can also be found in the seeds of flowering plants *Ipomoea* (morning glories), as well as *Argyreia nervosa* (the Hawaiian baby woodrose).[49] Why hallucinogenic chemicals should appear in such distantly related forms of life is not known.

The synthetic chemical ergotamine was packaged and marketed under the trade name Gynergen as a 'haemostatic remedy in obstetrics' and treatment for migraine. Sixty years later victims of cluster headaches[50] successfully campaigned for clinical trials[51] to investigate the potential use of LSD as a treatment for their affliction. This medical use has Depression-era roots: English pharmacologist Sir Henry Hallett Dale (1875–1968) discovered in the 1930s that ergotoxine has antagonistic effects on the production of adrenaline and the energetic volume of the autonomic nervous system (the involuntary component of our neural networks). Sir Henry believed it could be used in therapeutic ways.

Thus, Hoffman was charged with scrutinising ergotamine and its chemical relatives for their medical potential. He spent years playing in the laboratory, switching atoms around and coming up with new molecular variants. He made a profound contribution to the health of women, developing a new ergobasine 'that even surpassed the natural alkaloid in its therapeutic properties' he wrote[52]. This is today sold under the trade name Methergine, and is still one of the leading medicaments employed as a utertonic in obstetrics.

In 1938 he produced one molecule, LSD-25 (which, by the way, also has strong effects on the uteruses of laboratory animals). This one perked his interest:

> 'The research report also noted, in passing, that the experimental animals became restless during the narcosis. The new substance, however, aroused no special interest in our pharmacologists and physicians, and testing was therefore discontinued.'

The hallucinogen that would change the world then remained untouched on a laboratory shelf in the Swiss mountains for five years. In 1943 Hoffman returned to the molecule:

I could not forget the relatively uninteresting LSD-25. A peculiar presentiment – the feeling that this substance could possess properties other than those established in the first investigations – induced me, five years after the first synthesis, to produce LSD-25 once again ... during the purification and crystallisation of lysergic acid diethylamide in the form of a tartrate (tartaric acid salt).

His diary entry, 16 April 1943:

I was forced to interrupt my work in the laboratory in the middle of the afternoon and proceed home, being affected by a remarkable restlessness, combined with a slight dizziness. At home I lay down and sank into a not unpleasant intoxicated-like condition, characterised by an extremely stimulated imagination. In a dreamlike state, with eyes closed (I found the daylight to be unpleasantly glaring), I perceived an uninterrupted stream of fantastic pictures, extraordinary shapes with intense kaleidoscopic play of colours.

Charmed and intrigued, he was puzzled: 'Because of the known toxicity of ergot substances, I always maintained meticulously neat work habits. Possibly a bit of the LSD solution had contacted my fingertips... There seemed to be only one way of getting to the bottom of this. I decided on a self-experiment.'

Like many a gentlemanly and intrepid scientist, he decided to try the molecule on himself first.

Hoffman took 0.25 mg of lysergic acid diethylamide tartrate

34. An LSD blotter tab adorned with a depiction of Hoffman's first trip – possibly the most famous experience in the history of psychedelics. Suitably, the first LSD foray occurred during an actual journey in real space and time.

(diluted in 10 cc of water). His journal entry: 'Tasteless. 17:00: Beginning dizziness, feeling of anxiety, visual distortions, symptoms of paralysis, desire to laugh. Supplement of 4/21: Home by bicycle.' As he rode, he began to trip.

This psychedelic cycle ride is one of the most famous tales in the history of drugs. Interestingly, the world's 'first trip' was a bummer: Hoffman spent the next two days sick and miserable. 'A demon had invaded me, had taken possession of my body, mind and soul.' His neighbour brought him milk, 'but she was no longer Mrs R., but rather a malevolent insidious witch with a coloured mask'.

After the initial nightmare faded, Hoffman found himself in a world transformed: sounds became colours, colours became sounds, 'renewed life' flowed through him, and he spent the day in a vibrating, transcendent, synaesthetic utopia. Plus, he discovered what fans of LSD-25 regard to be one of its most redeeming qualities: 'No hangover.'

Hoffman, however, was no hippy. Quite the opposite – he was a true 'company man', working for Sandoz for almost his entire career. So powerful was his first LSD experience, and so fraught with psychological peril, he never imagined the molecule could wind up in the opium dens or hash clubs of cosmopolitan Europe. 'The last thing I could have expected was that this substance could ever find application as anything approaching a pleasure drug.' Brilliant he may have been, but in some ways endearingly naive nonetheless.

Still, he and his superiors at Sandoz unanimously agreed on the molecule's scientific remarkability. A flurry of excited experimentation by biologists, psychiatrists and zoologists followed. LSD-25 – so powerful and so strange – seemed set to change our understanding of the brain. Hoffman believed it could be for psychiatry 'what the microscope was for biology and the telescope for astronomy'. Animal experiments generated mixed results: mice were unafraid of cats.[53] Cats in turn were afraid of mice.[54] A cat nursed a mouse from its mammary gland.[55] An elephant died within minutes of a dose of 0.297 grams.[56] The variability in experimental results generated just as much confusion as excitement.

Psychiatrists, more interested in manifestations of the mind than mechanics of the brain, immediately saw the potential. The first self-experiment by a psychiatrist took place in 1947 by Professor Stoll himself at the University of Zurich. The trend went global, but Switzerland, as the birthplace of LSD-25, remained the centre of psychedelic research on human subjects. This truth is rendered even more remarkable considering the nation didn't give the vote to women until 1971.

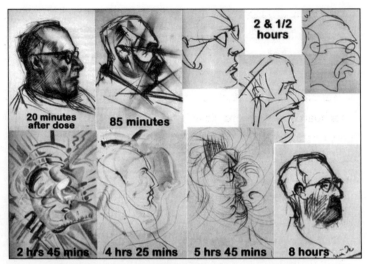

2 & 1/2 hours

20 minutes after dose

85 minutes

2 hrs 45 mins    4 hrs 25 mins    5 hrs 45 mins    8 hours

35. Experiments with LSD on the mentally ill?
Disturbing. Experiments with LSD on healthy artists?
If not illuminating, at least, highly entertaining.

The next two decades saw LSD widely deployed in the fields of psychotherapy, neurology, and every other form of biological scrutiny. Some psychiatrists thought LSD-25 could help them better understand and empathise with their patients. Shrinks could lose their heads for a day and return to reality with an enhanced appreciation for what it feels like for those who live permanently in the hinterlands of the mind. This was not a new idea: psychiatrists a hundred years earlier thought self-dosing with elephantine volumes of hashish would allow them to explore the realms of madness as tourists rather than residents.

Other psychiatrists thought LSD-25 could in fact *cure* madness, believing it might be the 'holy grail' for their profession: a test tube cure for schizophrenia. Institutionalised patients suffering the ravages of mental illness were given LSD-25.* Psycho-

---

* If you think this story is sad, it has nothing on the history of the prefrontal

therapeutic 'talking therapies' are expensive, time-consuming, and difficult. If a single-shot chemical cure could be found, psychiatric institutions could speed treatment by leaps and bounds. In other words: assembly-line psychiatry.

Patients were typically shut in a room with a teddy bear, a blackboard and a record player for entertainment.[57] It is interesting to note that the man who first created LSD-25 enjoyed this molecule in the sunny bucolic mountains of Switzerland, while the mentally unwell were confined in white square rooms to experience a drug that can be so unpleasant its victims have their own saint.

What did these psychiatric experiments teach us? Not a great deal. Far more interesting were the experiments conducted on artists, musicians and writers.

The most famous account ever penned of a psychedelic experience is almost certainly Aldous Huxley's *The Doors of Perception*, which became a sort of sacred text for teens experimenting with halucinogens. His insights could be deemed prophetic: as we shall see, his description of the brain as a 'reducing valve' has been validated by modern research.

## SILENCE OF THE SCIENCE

From an academic point of view, the most important person to get their hands on LSD-25 was the infamous psychologist 'Dr' Timothy Leary (1920–1996). His experiments on students and young offenders with acid during his time at Harvard University as a professor of psychology resulted in his dismissal and

---

lobotomy. More on this shortly.

arrest. More importantly, he ruined the chance anybody smarter than him may have had to legitimately explore the potential for LSD-25 to be used therapeutically for 40 years.

Following his expulsion, Leary launched into a highly successful career as a cheerleader for the mind-expanding properties of psychedelics. He is best known for his quip 'turn on, tune in, and drop out', and was described by President Richard Nixon as 'the most dangerous man in America'. For such a 'dangerous man', everything he ever wrote is unbearably boring.

Dull his prose may have been, but Leary's public persona was anything but. He aimed to turn American culture on its head and transform a generation with psychedelics. He unremittingly evangelised on the redemptive power of drugs with a number of unverifiable and outrageous claims (such as the assertion that LSD can result in 'thousands of orgasms' for women). Some kids had a great time. Others, not so much: some never came down. 'Acid casualties' still wander the streets of America.*

Hoffman, for his part, was horrified to see LSD abused by a generation of burnouts, and saddened to see it lead to so many lost minds. He felt that LSD came into a world that simply wasn't ready for it, and called it 'a wonder child that had to grow up in a dysfunctional society'.

Hoffman, however, did not burn out: he took LSD up to the ripe old age of 97 – and lived to be 102.

Following Leary's 'work' and the shutdown of credible science, almost all academic research with LSD could only be performed on animals. The results, as we have noted, brought mixed yet colourful results.

'From the bottom of the phylogenetic scale to the top, animals

---

* The most famous is the 'birdcage man' of San Francisco who, as his jacket description accurately envisions him, wears a birdcage on his head.

cannot effectively do other things when they are doing hallucino-gens,' wrote Ronald Siegel in 1989.

Clearly, Siegel was unfamiliar with the story of Dock Ellis.

Dock was a baseball pitcher in the 1970s who earned a number of impressive studs on his sporting belt. He helped lead the Yankees to win the World Series in 1976, was the starting pitcher for the National League in 1971, and totalled 1,136 strike-outs in nine years.* As a former baseball player myself, I'm most impressed by the fact that he was a switch hitter, meaning he could swing the bat from either the left or right side. This is like being ambidextrous, capable of writing with either hand.

Dock's career was decent but otherwise unremarkable, except for one thing: he pitched a no hitter. On acid.

'No hitters' are rare achievements. A pitcher must prevent all players on the opposing team from not only scoring a 'goal' – running round the diamond and 'scoring' at home base – but also from successfully hitting the ball and making it to the first base.

It's not an easy task. Many of the game's greatest pitchers have never managed a 'no hitter'.

But Dock did. On acid.

Now, it is not unusual for athletes to self-dose with contra-band chemicals: seemingly every week a star is caught cheating, goes on the telly, cries a bit, and then writes a bestselling book. Doping scandals are unremarkable.† Dock, however, was never caught. In fact, his teammates knew he took drugs with impres-sive regularity and let it slide because they knew he felt he needed them to perform.

Dock was heavily reliant upon simulants, particularly amphet-amines, Benzedrine and Dexamyl (a sprightly combination of

---

* For our British readers: that means he was pretty good.
† Caveat: Andre Agassi's crystal meth confession is rather noteworthy.

dextroamphetamine and amobarbital). He admitted, following his retirement, that he rarely pitched without their assistance. He said the scariest point in his entire career occurred when he had to play a game sober.

One day he was forced to pitch while high on LSD. He had been relaxing at his home in LA, took a few tabs, woke up and took some more, thinking it was Friday (the day before the game). Then he realised it was Saturday morning. (What happened to Friday? Nobody knows.) So he had to play. That afternoon. In another city. A man true to his duty, he hopped on a plane to San Diego.

Magically, he threw a no hitter. Which is particularly entertaining because he 'walked' a few batters, threw erratically into the stands, and ran away from the ball in terror. 'I jumped because I thought the ball was coming FAST … but it was coming SLOW.' He also struggled to recognise the ball's size. 'Sometimes I thought it was a BIG old ball … but then sometimes it would look SMALL.'

The story is instructive. Substances which one might think would hinder one's capacity to execute a complex task can actually render one more skilled than ever before. 'It was easier to pitch with the LSD because I was so used to medicating myself. That's how I was dealing with the fear of failure.'

Is his achievement an isolated tale? Not at all. Numerous physicists, geneticists, computer scientists and Nobel Prize winners have publicly stated that experimentation with mind-bending psychedelics catalyzed their insights and discoveries.

As for Dock, he later became a drugs counsellor and spent many years advising young people on substance dependency before dying in 2008.

# EXPERIMENTAL REVELATIONS

Dock was not alone in his addiction to chemical aids to do what he was best at. Countless musicians, artists and writers have become dependent upon stimulants, sedatives and hallucinogens in order to labour productively. The chronology from the biography *The Strange and Savage Life of Hunter S. Thompson*,[58] documenting a day in the life of HST, bears repeating:

*3 p.m – rise*
*3:05 – Chivas Regal with the morning papers, Dunhill cigarette*
*3:45 – cocaine*
*3:50 – another glass of Chivas, Dunhill*
*4:95 – first cup of coffee, Dunhill*
*4:15 – cocaine*
*4:16 – orange juice, Dunhill*
*4:30 – cocaine*
*4:54 – cocaine*
*5:05 – cocaine*
*5:11 – coffee, Dunhills*
*5:30 – more ice in the Chivas*
*5:45 – cocaine*
*6 p.m. – grass to take the edge off*
*7:05 – Woody Creek Tavern for lunch [beer, margaritas, cheeseburgers, fries, tomatoes, coleslaw, taco salad, onion rings, carrot cake, ice cream, more beer and Chivas etc.]*
*9 p.m. – cocaine*
*Midnight – Hunter ready to write*
*12:05–6 a.m. – Chartreuse, cocaine, grass, Chivas, coffee, Heineken, Clove cigarettes, grapefruit, Dunhills, orange juice, gin*

*6 a.m. – in the hot tub – champagne, Dove Bars, fettuccine*
*Alfredo*
*8 a.m. – Halcion*
*8:20 – sleep*

Most important detail: Midnight.

Scientists are equally vulnerable to the valley of chemical dependency for their creative and intellectual output. As we have noted, scientists are human, humans are flawed, and brilliant people still do stupid things. Scientists thus can become reliant upon stimulants to maintain energy levels, use alcohol and tobacco to cope with stress, and (this is where it gets interesting) achieve creative insights through the use of hallucinogens.

Mathematician Dr Ralph Abraham of the University of California in Santa Cruz has been described as a 'drug dealer to the geeks'.[59]

'I would describe myself much more as an encourager than a purveyor,' he says.[60] An ardent fan of enlightening hallucinogens, Dr Abraham disseminated them to California's scientific community in the 1970s hoping they might spark in his colleagues what they fired in his own mind. 'Psychedelics gave me the ability to look at computer graphic images and understand them in a way I just couldn't before. DMT in particular provided me with a gymnasium of visual skill.'[61] He says that without psychedelics he would not have advanced his understanding of chaos theory.[62]

Caveat: one does not need drugs to achieve profound insights into the structure of matter or the nature of mathematics. Benoît B. Mandelbrot (1924–2010), who illuminated a stunning series of fractal patterns including the eponymous Mandelbrot set, apparently never took psychedelics, despite the fact that he undoubtedly had ample opportunities. Sober he may have been, but his designs became the staple decoration of head shops worldwide.

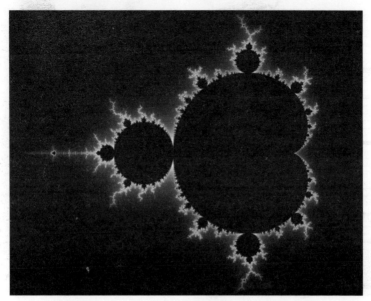

36. The famed Mandelbrot fractal pattern, adored by
hallucinogen-infused hippies, created by a sober-minded
mathematician. Sometimes the twain shall meet.

Physicists don't need psychedelics to achieve insights – but
they can help. In *How the Hippies Saved Physics* (2011),[63] author
David Kaiser tells the little-known story of how psychedelics
brought the moribund ideas of quantum mechanics back to life.

Imagine it's 1960, and thousands of kids with PhDs in physics
find themselves with nothing to do. During the Second World
War and the Cold War the US government aggressively funded
nuclear physics research. An army of clever geeks was enlisted
to outdo the Russians, and the focus was placed squarely on
the atom. Any department requiring funding had no time for
the airy-fairy ideas of quantum mechanics, first postulated in
the 1920s and 1930s by Erwin Schrödinger (1887–1961), Werner
Heisenberg (1901–1976), and Paul Dirac (1902–1984). The great
ideas that today are changing everything we know about what

37. *The Fundamental Fysiks Group.* You know the score: erratic hair, shiny buttons, wild eyes, Indian fabrics. Equation: Quantum Physics + New Age Ideals = This

makes up matter were left by the wayside, deemed too weird and insubstantial for serious research. Great ideas gathered dust on library shelves, forgotten and denigrated. But when the cold war thawed, funding for nuclear research melted away. And there were no jobs for tens of thousands of kids with PhDs in physics.

Suddenly they had time to read, think, and take drugs.

This injection of free time and the geographical convenience of being in new-age California brought them into contact with odd ideas, such as the potential for time travel, multiple universes, psychic communication and extra-sensory perception. Like author Ken Kesey's Merry Pranksters, they conferred on themselves a grandiose, wonky title: the 'Fundamental Fysiks Group'.

There were some strange results. A few serious physicists who had shown genuine promise went haywire, devoting their physical energy to hosting hot tub hallucinogen parties on Californian cliffsides.

But there were also some spectacular outcomes. Clever kids who finally had the time to relax after the gruelling ordeal of a PhD lounged around libraries, dusted off forgotten papers, and brought brilliant ideas back to life.

Result: quantum physics entered the world of serious academic research once more. Was hippy counterculture partially responsible for the quantum profusion we enjoy today? Just maybe.

Computer science too has been altered by mind-altering substances. One of the most famous quips comes from Steve Jobs, who described taking LSD-25 as 'one of the two or three most important things I have done in my life'.

Jobs may have publicly trumpeted his supposed use of psychedelics, but it is more contentiously claimed that molecular biologist Professor Francis Crick (1916–2004) deduced the structure of DNA via the insights gleaned by the use of LSD-25. Crick, along with James Watson, deduced the structure of deoxyribonucleic acid, or DNA, the spiralling chemical language of life.

The Nobel Prize in physiology or medicine in 1962 was awarded to the two molecular biologists plus the head of the x-ray department at King's College London, three winners in all. (Or was that four? But hey, who's keeping count. See: Rosalind Franklin, unfortunately a mere footnote in the history of science, the woman who did much of the work and received none of the credit.)

Crick disputed that LSD gave him an intellectual leg-up, though he purportedly told the *Daily Mail* reporter who leaked the story: 'Print a word and I'll sue.'[64]

More proud of his hallucinogen-assisted discoveries is fellow Nobel Prize winner Kary Mullis, who devised the polymerase chain reaction, also known as PCR, a genetic unravelling worthy of awe.

Geneticists in the 1980s were faced with what seemed to be an insurmountable challenge: how to read the entire genetic instruction manual of a species, the 'genome'. The string of chemical building blocks that comprise the total amount of DNA is difficult to conceive. If you imagine this genetic instruction manual as a book, a human's book contains 3.2 billion 'letters'.[65]

The language of life is universal: it unites every living thing, from bacteria to cane toads, bonobos and fungi. We all use the same code. This language has four letters. Each letter is known as a 'nucleotide': adenosine, thymine, cytosine and guanine, to be precise, shorthanded as A, T, C and G.

A strand of DNA can read something like this: ATCTCTCTGGGAATTCGAGAGCTAT.

Reading this out, one letter at a time, was a glacially slow, frustrating process for patient geneticists in the 1980s.

Mullis devised an ingenious way to shatter this multibillion letter spiralling string into a splattering of shards that could be read and then reassembled back together.

His insightful innovation: build chemical robots (dubbed 'primers') that recognise a specific string of letters – such as GGGAATTCGAGAG – and from there read out the recipe. So imagine a chemical robot sees the word 'yeast' or 'roast' or 'mix'. Every time that word crops up, the robot reads the recipe and prints a copy. Perform this over and over, and instead of a 3.2 billion letter long string of A, C, T and G, you produce a bunch of chunks which are manageable to read – in other words, extracts rather than an entire book. It changed forever the face of genetics, and garnered Mullis the Nobel Prize in 1993.

In his autobiography *Dancing Naked In The Mind Field* (2000),[66] Mullis describes how LSD helped reveal the answer to the genetic puzzle he was trying to solve: 'DNA chains coiled and floated. Lurid blue and pink images of electric molecules injected themselves somewhere between the mountain road and my eyes.'

Sorry. What's that?

*Mountain road?*

Mullis achieved his crucial insight on acid. While driving. His mode of discovery may have been unorthodox, but the importance of his achievement cannot be understated. The Nobel committee certainly thought so. The procedure he devised through the origami of his addled brain revolutionised our understanding of the language of life.

But let's not label Kary Mullis an intellectual or psychedelic hero. He has denied the scientific evidence for climate change and – wait for it – AIDS.


# RENEGADE RESEARCH

No discussion of scientists or drugs would be complete without Dr Alexander Shulgin, the man who gave the world MDMA, descriptively known by its more common name: Ecstasy.

After obtaining his PhD from the University of California, Berkeley, Shulgin worked for the Dow Chemical Company throughout the 1960s developing pesticides. His most profitable creation was Zectran, one of the world's first biodegradable insecticides. The proceeds from this insect repellent were considerable, and allowed him the freedom to pursue his own rogue interests.

A colleague introduced him to a curious compound, 3,4-methylenedioxy-N-methamphetamine, first synthesised in the laboratories of pharmaceutical giant Merck in Germany in 1912. A few animal experiments had produced uninteresting results. Nobody

38. Alexander Shulgin (1925–2014). He gave the world MDMA, crafted hundreds of drugs in his shed, and celebrated the art of chemistry in a way never seen before. RIP you irreplaceable renegade researcher.

knew much about it. A few scientists had investigated the potential for the strange molecule to be used as an antispasmodic, but little came out of their studies.

Somehow the chemical wound its way into American subculture, most likely due to its deployment by the American military on Canadian psychiatric patients as part of the MKUltra programme. These notorious experiments in the 1950s by the CIA sought to perfect the art of 'depatterning' interrogation subjects: break them to remake them.

Like mescaline and LSD-25 before it, rebellious twentieth-century counterculture was introduced to this chemical thanks to military attempts to employ it as an agent of control. And like LSD before it, somehow, the chemical leaked. By the 1970s, rogue underground experimentalists were dabbling with MDMA. The first recorded seizure by police took place in Chicago in 1970.[67]

But it hadn't hit the big time. Few knew about it – or, more importantly, how to *make* it.

In 1976, Shulgin was introduced to the chemical by a student, who reported that it had remarkable effects. In the spirit of discovery, Shulgin ingested it. He would later replicate this method, always trying every single chemical he concocted on himself first.* His first MDMA experience was transformative. Shulgin became obsessed with the capacity for synthetic compounds to transform human experience, and began to focus his research into their synthesis, structure and effects. He worked for Dow for a few more years, and published reports on his chemical creations in reputable journals such as *Nature*[68] and the *Journal of Organic Chemistry*. When Dow grew weary of his druggy pursuits, he set up shop in his garden shed.

Shulgin single-handedly crafted thousands of chemicals, of which hundreds were found to be psychoactive. A kooky chemist as well as gentleman, he always tested every creation on himself first. Then, if it seemed safe, his wife. Then, his friends. Then he would distribute the chemical formula to the public through his website Erowid.[69] In addition to decoding and broadcasting MDMA to the world, he created the club drugs 2C-B ($C_{10}H_{14}BrNO_2$), 2C-E ($C_{12}H_{19}NO_2$), and a few that behave in weird ways, such as diisopropyltryptamine ($C_{16}H_{24}N_2$) which only influences the auditory system: DIPT is a hallucinogen for the ears.

Shulgin consistently evaded authoritative measures because everything he synthesised was new and therefore uncontrolled. Every time he made and took something he had created, his consumption was therefore entirely legal.

Fearing he would eventually be shut down by the DEA in

---

* One word best describes such behaviour: chivalrous.

America, Shulgin self-published his complete chemical compendium – the tale of each molecule's discovery, their effects and how to produce them – in a two-volume chemical bible: *PiHKAL* (1991) short for *Phenethylamines I Have Known and Loved* and *TiHKAL* (1997), short for *Tryptamines I Have Known and Loved*. Both are part chemical textbook, part autobiography, each tome is 800 pages long.

Shulgin is regarded as a hero by some and a menace by others.

An entire generation of ravers would say MDMA set the tone for their twenties, and changed their lives for the better. But not everyone had a great time with his chemical concoctions. In 1967 2,5-dimethoxy-4-methylamphetamine, DOM ($C_{12}H_{19}NO_2$), became briefly fashionable in the San Francisco hippy enclave of Haight-Ashbury. A few kids found themselves in hospital thinking they would never come down.[70]

Ecstasy was Shulgin's Number One Hit: throughout the 1980s and 1990s it spread across the planet. As the chemical made an inexorable march through subcultures and into the mainstream, parents worried, authorities disapproved, and newspapers screamed.

Just how dangerous is MDMA? Initial studies in monkeys indicated the compound could result in neurodegeneration[71] (permanent deterioration of the electrical highways of the brain). But these have failed to manifest in a generation of human MDMA users, 25 years later. One study indicated the chemical could trigger the symptoms of Parkinson's – but that turned out to be the result of mislabelled lab samples.[72] The compound was in fact methamphetamine, or crystal meth.

Many of the deleterious impacts of MDMA may simply be due to its combined use with alcohol and other drugs. An intriguing study of Mormon teenagers[73] who had access to MDMA but not alcohol showed no discernible difference in their mental functioning compared with teenagers who had never taken drugs. But due to the clampdown on research following the American DEA's restriction in 1985, MDMA was employed in only a few small studies on depression and anxiety. Neuroscientists could not investigate how the compound affects the human brain.

It is not only licenses and red tape which make clinical studies of psychotropic drugs so difficult: the expenses are enormous. How much do you think one gram of medical grade MDMA costs? £500? £1,000? Answer: £5,000. For one gram. Considering you can obtain the compound for twenty quid on the street, it's remarkably high, and makes bona fide research prohibitively expensive. 'The ultimate effect of all these rings of red tape is to harm the science,' says Dr Robin Carhart-Harris of Imperial College in London, who conducted one of the world's first studies of psilocybin on the human brain using MRI scanners.

'Science is hard: it requires special licenses and regulations, and most people just don't have the patience for it – it is utterly tedious sometimes,' says Dr David Nutt, Director of the Neuro-psychopharmacology Unit at Imperial College London. You may know him as the man fired by the British government for statistically demonstrating that taking ecstasy is less dangerous than horse riding.*

Professor Nutt's use of magnetic fMRI scans revealed patterns of activity induced by MDMA that had not been seen before: a decrease in the 'synchronous firing' of two regions, the prefrontal cortex and the posterior cingulate cortex.

In people who suffer from clinical depression, these two

---

* One wonders if he would have caused such a stir if he had highlighted the risks of skateboarding or sky-diving instead of singling out equestrian activities. Was it the choice of an upper-class pastime that ruffled their feathers so?

locations are known to synchronise their firing more tightly, which is thought by many neuroscientists to correlate with 'ruminative' thought patterns.[74] Spinning round and round, dwelling on negative ideas, memories and feelings. Trapped in an emotional vortex. You've probably been there.

## RECLAIM THE DREAM

A popular drug among contemporary ravers: $(RS)$-2-(2-chlorophenyl)-2-(methylamino)cyclohexanone, also known as ketamine, horse tranquiliser, or – the most apt description – 'regretamine'. Chemical formula, $C_{13}H_{16}ClNO$. The unusual inclusion of chlorine – a component of bleach – in this molecule partly explains the destructive impacts it has on nasal passages, teeth and bladders.

Most tranquilisers and dissociatives, such as benzodiazepines, barbiturates and alcohol, act upon the 'GABA' receptor system: they lock into the docking bay designed for the neurotransmitter 'gamma-aminobutyric acid'. But ketamine, like its weirdo cousin phencyclidine (PCP, also known as 'angel dust'), works via the glutamate receptor system. It's odd.

Ketamine is neither an upper nor a downer, it's a 'sideways'. Cheap and easy to acquire or manufacture, it is easily found wherever there is loud music.

Although he's 'never tried it', psychiatrist Dr Adam Winstock, Honorary Consultant and Addictions Psychiatrist at Maudsley Hospital and Lewisham Drug and Alcohol Service, says he can appreciate why some people dig it.

'I can understand the appeal of being entirely removed from yourself in space and time,' he says. 'If you look at the statistics from Hong Kong, the number one drug of abuse is ketamine. Countless kids are born to imported China brides married to older businessmen. Kids are brought up in loveless households, besieged by domestic violence. If you were in that situation, of course you would want to be disassociated.'[75]

Winstock is also the founder of the Global Drugs Survey and Drugsmeter.com, an online repository of recreational drug data, where you too can plug in your habits, experiences and preferences. In return for sharing anonymous data, you receive 'objective, personalised feedback that takes your personal features in to account'.[76]

Why give your data away to learn about people who take the drugs you do?

'I am concerned that people who use drugs and like to take drugs don't have readily accessible information that is presented to them in a way that is acceptable, personally meaningful, or relevant,' says Dr Winstock. 'The truth is that most people are pretty smart, like to keep themselves safe, but still tend to take drugs and alcohol in a way that puts themselves at risk. And they know it. Probably due to the tendency of human nature to overestimate our invulnerability to risk.'

So if smart people do stupid things, what's the point in gathering information and spreading it online?

'Individuals are not defined by their drug use. People who take drugs recreationally also go to the gym, have jobs, are normal people,' he says. Dr Winstock, it should be noted, claims he has never taken MDMA, LSD, or any other narcotic himself. He asserts he was drawn to the field after observing the behaviour of his friends when he was a university student. 'I LOVE YOU MAN. NO, YOU DON'T UNDERSTAND. I LOVE YOU.' The cognitive impacts proved too profound to ignore.

'We need to communicate that the vast majority of people's

lives haven't been ruined by drugs,' he says. 'It's about when you take them, and what you take. There are 101 things that drugs do for different people at different times. We should be aware of which ones are the most harmful, or most desired. Frankly, most people won't want crack or PCP if they can get other things. If you have LSD, MDMA, cocaine and cannabis, you don't really need anything else.'

But when these are in short supply, people turn to other drugs. Sometimes, in search for 'treatments for the human condition', people dabble in, delight in, and become dependent upon compounds that are unsavory, lethal – and legal.

If you ask people 'what is the worst drug you have ever tried?' I find there are typically two answers.

One: *Salvia divinorum*. Also known as seers sage and Jimson weed, salvia has not only been tested as a treatment for depression,[77] it remains legal in most western countries. If you've ever smoked the raw leaf, you probably were unimpressed, as it is far less potent than marijuana. However, chemists have developed ways to refine the plant into pellets, which are typically sold in 10x and 40x concentrations. They are extremely strong. And they are legal.

The other: $C_2H_6O$, ethyl alcohol, also known as ethanol, or simply, booze.

It is unlikely that any substance has been so widely deployed in the battle between heart and mind. Perhaps the Bard put it best in *Othello*: 'Oh, that men should put an enemy in their mouths to steal away their brains!'

Alcohol might seem like an unremarkable drug, compared to otherworldly hallucinogens such as LSD or exhilarating stimulants such as cocaine. But in fact, it is quite interesting from a biochemical point of view, which may shed light upon its ubiquitous popularity. Alcohol ticks many neurochemical boxes, which goes some way to explaining why it is one of the most universally popular drugs across the animal kingdom: unlike nicotine, which is lethal to nearly all living things, booze is one of evolution's greatest intoxicating hits. Monkeys, elephants, dogs, insects, pigeons, and just about every lab animal subjected to forced ethanol intake take to it like ducks to water.

It stimulates not just one, or two, but a bouquet of receptors. Alcohol locks into the keyholes for our neurotransmitter acetylcholine, which we encountered as the target for the nicotine in tobacco and the muscarine in fly agaric mushrooms. It invades the cellular cradle for the native neurotransmitter glutamate, which the quirky psychedelics ketamine and angel dust prey upon. Alcohol also moors into the harbours of serotonin's bays, which already generously accept the hallucinogens LSD and psilocybin with open arms. And like barbiturates and other sedatives, alcohol snuggles into the inviting nests designed for the neurotransmitter GABA.

Alcohol is impressively promiscuous in its capacity to cosy up to a variety of our biological receptors. However, it is also uniquely poisonous: known as the 'dirty drug', the toxic solvent has the ability to roam far and wide, infiltrating our cells from head to toe. Bad hangovers come with widespread muscle aches for a reason. The insidious drug is a nimble chemical ninja. Its disastrous effects on the liver are well known, but it also has profound impacts on the cerebellum, the wrinkly bulb at the back of the brain, central for balance and motor control. Hence booze's signature wobbles, stumbles, falls and fatal car crashes.

Few nations have fully embraced the use of alcohol as a medicament for the human condition than the UK. Newspapers daily mourn the 'binge drinking' culture of 'boozy Britain', but this is not a new public health hazard. Historic records indicate that the Brits of centuries past lived in a permanent alcoholic haze. Stalls in London in the time of Samuel Johnson (1709–1784) advertised,

'Drunk for a penny, dead drunk for twopence'. The alcoholic fibres of European society, however, stem from practical roots: cider and beer were usually safer than drinking water due to alcohol's antiseptic properties.

Alcohol continues as Britain's drug of choice. It is ethanol more than any other that Professor Nutt says we have neglected to consider in a scientific manner.

'The drinks industry has convinced the public that alcohol is not a drug,' he says. The death of singer Amy Jade Winehouse (1983–2011) proves that not only is alcohol a drug, it is dangerous. Post-mortem analysis proved the singer only had alcohol in her system at the time of her death. No heroin. No crack. Just booze.

Professor Nutt points out that there are 15 ways alcohol can kill you, from cirrhosis of the liver to cancer of the oesophagus. Yet in only one way is it beneficial, which is the well-known and yet marginal influence on cardiac health. To achieve this effect, one needs to limit one's intake to half a unit a day. Exceed this, and that benefit disappears. 'You have to hand it to the drinks industry,' compliments Dr Nutt. 'They managed to turn "threshold limits" into "weekly allowances". It was a brilliant stroke of marketing. We should learn from them how to communicate our own messages.'


## CURES WHAT AILS YOU

My answer to the question 'What is the worst drug that you have ever tried?' is Lamotrigine, a molecule that came from the laboratories of a pharmaceutical factory, handed over to me by a physician with a PhD.

I have temporal lobe epilepsy and in 2009 experienced a weird resurgence in my episodes after seven years without a single spell. During my university years, when I had subsisted off barrels of caffeine during the day and then buried my brain in potent Canadian cannabis in the evenings, I didn't have one seizure for five years. Then at the age of 27 when I slept normal hours and no longer had a taste for weed, I oddly started to have seizures again. This might actually be attributable to the ganja: there is ample evidence that marijuana can have an ameliorating effect on epilepiform episodes, and many epileptics smoke considerable volumes of marijuana for this very reason. Shame I don't like weed anymore.

After numerous tests in 2009 I was prescribed Lamotrigine – a drug initially developed for bipolar disorder – and against my better judgement, I took the prescription. Having taken all manner of dodgy chemicals on the floors of weirdo warehouse raves, it seemed unscientific to spurn a drug prescribed by a medical doctor with more academic credentials than myself.

Long story short: it destroyed my happiness and my health. I endured raucous nightmares every single night. And every single morning I woke up with limbs made of concrete. Seven months later I was two stone heavier, slept 18 hours a day, never went out, and only ever thought about sleeping. I can count on one hand the number of things I have done in my life that I regret, and taking that foul prescription tops the list.

The name 'Lamotrigine' still makes my stomach turn, but there are people who would be dead were it not for drugs like it. Conditions such as bipolar disorder are not to be discussed frivolously, nor are the drugs that pharmaceutical giants have developed to combat them.

There are many comprehensive surveys of the power, influence, scientific importance and cultural significance of the pharmaceutical sector. Medical doctor and journalist Ben Goldacre argues that pharmaceutical companies spend twice as much on marketing and advertising as they do on research and the development of new drugs, industry funded trials are twenty times more likely to give results favouring the test drug, and half of all clinical trials go unpublished.[78]

He catalogues the questionable recruitment practices for trial participants (in the jargon of the profession, 'paid volunteers', which he rightfully terms an oxymoron). This ranges from homeless people, prisoners and alcoholics, right back to the testing on conscientious objectors in the 'Great Starvation Experiment' (a series of American experiments, now considered controversial and ethically questionable, which saw study subjects starved to better understand the conditions allied soldiers and concentration camp survivors would be subjected to). Gifts and freebies for doctors, ghostwritten papers, the 'statistical poison' of missing data, the pathologisation of quotidian human experience, such as reclassifying the sadness that follows the death of a family member as 'adjustment disorder'.

Psychiatrist Dr Adam Winstock is also not a fan of the pharmaceutical industry. 'The single biggest source of death by drugs in America is doctors and pharmaceutical companies – more people die from doctors than any other source.'[79]

Dr Nutt does not, in fact, share Goldacre's view: 'Ben's got the wrong end of the stick on this one,' he says. 'Pharmaceutical trials are some of the most tightly regulated, stringently monitored tests that are conducted in all of medicine. They are extraordinarily well monitored.'[80]

Really? Well, Nutt has his own reasons for believing in their virtues: he is eager to partner with them and acquire funds for his own research.

'I'm a prostitute,' he says, jokingly but honestly. 'I will partner with whoever will fund my work.' Anyone who has slogged through the agonising hoop-jumping of a grant application can sympathise with his view.

So is this what we really want: pharmaceutically manufactured government-approved hallucinogens? I'm not so sure. On the one hand, pharmaceutical pills such as Valium are reliable in their quality compared with the gritty Russian-roulette of street-grade-anything. But on the other hand ... can we really trust Big Pharma? I don't. And I for one have good reason not to.

## ADDICTED TO ADDICTION

'Nothing in biology makes sense except in the light of evolution,' wrote Russian biologist Theodosius Dobzhansky (1900–1975) in 1973. He was a Christian, but firmly believed that species change over time: they evolve. If we deny that living things transform into new ones, we will never understand life on Earth in any way, shape or form. We must understand the past to comprehend the present.

Few things in science can be termed 'factual'. What we consider to be 'the truth' shifts over time, whether we are discussing the shape of the solar system, the structure of the atom or the difference between the sexes.

But biologists unanimously agree: evolution is a fact. How it *operates* is a matter of debate: natural selection is not a fact but a theory (albeit a pretty damned solid one).

Many evolutionary biologists once took as an article of faith that evolutionary change proceeds as an inexorable march towards 'progress'. The weak are left behind, the fit survive, leave behind more offspring and win the reproductive race. Bigger, faster, stronger. Onwards and upwards. As the Beatles would have put it:

'things are getting better all the time'. The term for this ideological framework: 'adaptationism'. The concept: almost everything we see in living things today is the optimal product of evolution.

But more careful scrutiny of the tree of life reveals countless examples that disprove this assumption: birds become flightless, cave-dwelling animals sightless, lizards legless. Frequently creatures inch backwards. There is no need to waste precious energy on maintaining an expensive piece of anatomical kit if it is unnecessary. Moreover, many features continue to exist not because they are beneficial but because they are neutral. They are not lethal, so their presence does not affect the survival of those who bear them. Wisdom teeth and body hair in humans are often cited as such 'vestigial organs': neither adaptive nor detrimental, they remain in place simply because natural selection cannot be bothered to get rid of them.

Other qualities – both beneficial and deleterious – survive because they are the by-products of a different and vitally adaptive feature. Sickle-cell anaemia is a classic example: people born with this condition bear blood cells that are twisted rather than round, and thus less able to soak up oxygen. One in every 500 African Americans are hobbled by this condition, including singer and spectacularly fit dancer Tionne Watkins (T-Boz) of 1990s R 'n' B group TLC, and Dock Ellis, the unparalleled baseball player who pitched the world's only no-hitter on LSD.

The dancer and the athlete both demonstrate that we are not always slaves to our biology.

Why should such a crummy condition be so common in America? Because sickle-shaped blood cells are more resistant to malaria infection than garden-variety blood cells. In Africa – where the lethal parasite remains a ubiquitous killer – twisted blood cells can actually be advantageous.

All this academic dithering is to say that just because a

39. Most of the lifelong addicts registered with Correlation
– a 'harm reduction' organisation in the Netherlands – are
hooked on both heroin and crack. By the bye: crack has a
distinctive smell. Three words: burnt rubber oatmeal.

seemingly beneficial trait exists doesn't mean it is evolutionarily adaptive. Similarly, a seemingly lethal trait may not be
maladaptive.

Why, then, is the human species so prone to the ravages of
addiction? If nothing makes sense except in the light of evolution,
why are we so easily prone to destroy ourselves with toxic drugs?
Addiction is one of biology's greatest puzzles.

Humans do many silly things, from life-threatening extreme
sports to mind-numbing television bingeing. But few things
could be as senseless as drug addiction: livers are destroyed,
teeth eroded, septums disintegrated, intellects eliminated, and
countless lives lost. And for what? A momentary high, followed
by a catastrophic low. Socioeconomic conditions are not the only
cause: addiction ravages not just the poor and the desperate, but
also the beautiful and the successful. You know the score: Janis

Joplin. Jimi Hendrix. Jim Morrison. Countless lost talents litter humanity's floor.

Few facets of the human condition could be less adaptive. So why is addiction so pervasive? One can read as many academic papers unpicking the neurochemistry of dopamine release as one likes. But nothing can inform your understanding of what it means to rely upon a toxic chemical for daily existence than personal experience.

Some years ago, as an employee of the European Commission, I was sent to Amsterdam to interview scientists at Correlation, a shining star of 'harm reduction' strategies in the EU's efforts to halt the spread of HIV.[81]

A beautiful medieval building housed a safe 'consumption room' – also known as a 'shooting range' in the parlance of junkies. A place where heroin addicts can inject with clean equipment.[82]

Upstairs: a row of clean white canteen tables. Fifteen junkies on each side chatting, burning, stirring, smoking and shooting.

'Hepatitis C is a terrible disease,' sighed the institution's director Dr Eberhard Schatz.[83] 'It eats away at you slowly and is horrible to watch. But the public doesn't know or care about it as much as they do about HIV, because you only get it from direct blood transmission. So it's mostly junkies who are affected.'

Clients can come as much as they like, from 8 a.m. to 11 p.m., to use sterilised equipment. The doorbell rings every time one of the registered users wishes to access the room. The bell rings constantly.

'We have no delusions that we can get most lifelong addicts off heroin or crack. Either they come here and they do it with needles we know are safe, or they do it on the streets,' Dr Schatz said. 'The honest truth is that they hate methadone. It numbs them, and in fact the withdrawal symptoms are far worse than

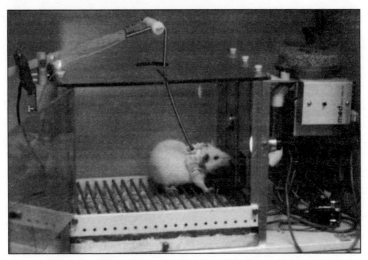

40. A rat with an apparatus for the self-administration of drugs half a century ago. If you had a metal tube implanted in your spine, would you have any reason not to take drugs?

those from heroin itself. It would actually be even more effective if we just gave them heroin like they do in Portugal, rather than letting them source the material from the street, where it could be laced with anything.'

Treating addiction with the object of their addiction.

What, exactly, does it mean to be addicted? It's a question that has generated a surprising variety of answers. In the nineteenth century it was largely seen as a form of 'moral weakness'. Most people, it was reasoned, should have the capacity to consume opium or cocaine on occasion without becoming hooked.

But as neuroscientists began to unravel the nature of receptors and the ways drugs 'hijack' our body's machinery, the chemicals themselves became the target of blame.

The chemical markers of addiction seem to come in two flavours. One is 'sensitisation', in which the brain's capacity to latch on to narcotics is ramped up. This seems to be the culprit at play behind the intense cravings addicts suffer from, and why rates of relapse can be so high. The other is 'tolerance', when users require higher and ever higher volumes of their chosen poison to achieve the same high. This grim chemical reaper is usually responsible for overdoses, which are more often than not accidental; habitual users overestimate how much they can handle.

Neurological evidence that drugs induce measurable changes in the brain piled up, and the 'addiction as disease' model now prevails. The concept seems to be validated by decades of laboratory work on animals. Monkeys who are taught to press a lever to receive a cocaine reward will do so voraciously – even if the number of lever presses required to deliver the hit is systematically increased. If they have to press the lever 100 times for a coke injection, they will do so; if this requirement is increased to 200 times, they will still hammer away for a hit. Monkeys have been known to relentlessly jab at the trigger more than 12,000 times for the deliverance of a dose.[84]

The 'addiction as disease' ideological framework posits that narcotic substances are themselves the problem, and those who take them suffer from a medical problem. Drugs are poison. Addiction, in this light, is seen not as a moral weakness, but as an illness. On the upside, it implies that addicts are not criminals, but unhealthy individuals worthy of compassion rather than incarceration.

The downside: we are slaves to our biology.

Let us recall the forgotten work of scientist Bruce Alexander of Canada's Simon Fraser University, a stone's throw from Vancouver, a coastal town renowned for vistas of mountains and trees, and slums teeming with junkies. In the 1970s Alexander suspected lab conditions were influencing the behaviour of animals. Work by other researchers had suggested that laboratory animals, if given the chance to 'self-inject', will do so to the point of self-obliteration; in one study on lab rats implanted with needles in their spines, 83 per cent of the animals became 'heavy users', and after a month, 90 per cent of the animals had died.[85]

But, reasoned Alexander, sequestering rats into cages with needles implanted into their spines so they can 'self-inject' with morphine or cocaine may not realistically replicate normal rodent living conditions. Stressed animals may skew the data with their desire to self-medicate.

Most readers will probably recognise in this the obvious parallel: prison. It is easy to acquire heroin and crack within prisons. This is exactly where many people try hardcore drugs for the first time.

Alexander devised an alternative scenario for his rats: rather than tiny cages, why not create a comfortable setting? He constructed an innovative habitat for the rats, nicknamed today 'Rat Park'. Instead of cages, devoid of company and comfort, the rats were given ample space, grass for bedding, other rats for fornicating, and a variety of nibbles, wheels and toys. His grad students even painted murals of boreal landscapes to complete the setting.

He then made the rats addicted to opiates through forced morphine injections. Then, instead of leaving them to self-inject in a tiny cage, where many of their rodent counterparts died from overdose, he transplanted them to Rat Park.

When given the choice between morphine-laced water and regular water, many of the addicted rats opted for plain water, consuming a twentieth the amount of opiates as addicted rats confined in cages. When they could play and copulate, drugs became less appealing.[86]

Implication? Drug use and dependence is exacerbated by socio-economic conditions. Poverty and loneliness predispose vulnerable people towards narcotic use. Not a controversial statement. But Alexander's funding was pulled a couple of years later, and as a result his work is largely forgotten. So it goes.

Does this mean that if most humans have a decent standard of living and the freedom to eat, play and fornicate as they wish, chemical addiction would vanish from our world? It's an appealing idea.* But scientific modes of thinking force us to remember that the more seductive or appealing the idea, the more worthy it is of scrutiny.

Recent studies provide more clues as to why people destroy themselves with addictive chemicals.

Imagine a vista with ample space, lots of food, and plentiful company for copulation: a world saturated in space, sex, and sugar. The modern western world, perhaps? America, Canada, Europe, and increasingly the world's fastest developing nations (such as Brazil, India and China) suffer from elevated levels of obesity, diabetes and hypertension due to high levels of sugar consumption.

You might be thinking… 'Sugar? Sugar isn't a drug. It's a food.'

Many things would lead any scientist to investigate the addictive and cognition-altering qualities of sugar – base instinct being one of them. Take the glint in the eye of a child clamouring for candy. It is remarkably similar to that of a crack addict.

Careful scrutiny of the chemical messengers of the brain indicates that sugar can behave much like a drug.

'There are overlapping neural circuits between food intake and substances of abuse – when people overeat palatable foods, the brain systems look more like they do as a drug of abuse than you see with simple casual eating,' says Dr Nicole Avena of the Department of Psychiatry at the University of Florida. 'Overconsumption of sugar can have addictive effects on the brain.'[87]

Her research has shown that sugar and fat-rich foods – or

---

* Though not a realistic one if we consider the lifestyles of celebrities.

'hyperpalatable' foods as she terms them – can release dopamine in levels and patterns of receptor activation akin to what is seen in rats dosed with morphine or cocaine. These scientific results reinforce what we anecdotally observe in human beings. Psychologists call it 'addiction transfer': swapping one fix for another. Give up booze, devote oneself to God. Ditch cigarettes, eat voraciously and pack on the pounds. Abandon the needle, switch to an inexplicable and unprecedented love for roulette. Round it goes.

It is enticing, says Avena, that we see similar patterns of dopamine receptor activation with crack cocaine and chocolate cake. Yet this is a small slice of the story. Neuroscientists are limited by the number of 'ligands' that they can work with. Ligands are molecules that bind to a specific neurotransmitter which are designed for laboratory monitors to detect and trace. Like shining headlights, they help us map the highways of the mind.

Ligands allow us to see what signal goes where. We have strong and reliable ligands for dopamine, but not for most other neurotransmitters. Dopamine is a celebrity among chemical messengers, but it is just one of hundreds.

'We need to go "beyond dopamine"', Dr Avena explains. 'It's useful to simplify things so the average person can have a rudimentary understanding, but it's important to remember that there are interactions we need to understand in more detail to truly understand addiction. What are the mediating systems that affect dopamine? What role do they play? We will better understand things as the technology improves and we develop stronger ligands, but at the moment we are restricted to dopamine.'

Whether you care about the fate of addicts or not, from a neuroscientific point of view, addiction is worth investigating, says Dr George Koob, Chair of the Committee on the Neurobiology of Addictive Disorders at the Scripps Institute in California.[88]

Addiction teaches us a lot about how the brain works. In addictive patterns, he says, we erode the body's natural systems for administering rewards, and in exchange it cranks up the stress circuits.

'The brain becomes activated and stays activated as a response to excessive consumption. There is an important part of our reptile brain that is involved in not feeling bad – how that system changes is an important component of addiction,' he says. 'Understanding addiction helps us to understand how our emotional system functions.'

Or as Avena puts it, addiction is just the other side of an important *Homo sapiens* characteristic: learning.

'Addiction is about learning to do something really well,' she explains. In other words, if something feels good – like sex, or eating, or sleeping – having the capacity to learn to do it again and again is good for you. Individuals that learn to do pleasurable things well will be more likely to survive, reproduce, and pass on their hedonistic genes.

Narcotic drugs like cocaine result in the release of pleasurable neurotransmitters like dopamine, so our brains are unfortunately primed to want to take them again and again. Crack – which packs a bigger punch – is thus more addictive and more dangerous because it delivers a more potent burst of dopamine.

'We think of dopamine as being associated with reward, but it actually has a lot to do with learning,' says Dr Avena.

In other words, our capacity to think, ponder and learn – 'higher' cognitive characteristics we consider to be central to what makes us human – are the very reason why we are prone to destroy ourselves with corrosive, erosive and soul-destroying narcotics.

Today, most legitimate research on illegal drugs, including LSD, psilocybin and MDMA (all 'Class A' drugs in the UK, classified in the same category as crack and heroin), is not as intoxicants but as medicaments. Rather than a means to addle the mind, clinicians, neuroscientists and psychologists are investigating the potential to achieve balanced mental states using heavy narcotics. This harkens back to the middle of the twentieth century, when LSD was investigated as a treatment for schizophrenia and a litany of mental ailments.

If the study of evolution has taught us anything, it is that the complex brains that make humans so special also render us vulnerable to ailments that make the very point of living questionable. Our brains make our world. And sometimes our brain is our enemy.

Depression remains one of our most intractable afflictions, and it is why so many people get into drugs in the first place. Alcoholism remains ubiquitous: it is notoriously difficult to treat and cursed by high rates of relapse.

Which makes one of the most potent cures which clinicians and pharmacologists are now investigating even more intriguing: LSD.

Countless alcoholics have anecdotally reported that a psychedelic conversion changed their lives. Bill Wilson, the founder of Alcoholics Anonymous, is a particularly famous example, and an interesting one given the association's Christian template. That an organisation which demands its followers admit they are 'helpless' and adhere to religious doctrines was founded through a narcotic conversion is noteworthy.

An unfortunate case is that of rogue American chemist Casey

Hardison, who in 2004 was sentenced to 20 years imprisonment for mass manufacturing and distribution of LSD. From prison, Hardison has written extensively on his punishment, and defended his actions. 'LSD gave me a rare glimpse of the power of the human mind to shape reality,' he writes. 'The so-called war on drugs is not a war on pills, powders, plants and potions, it is a war on mental states, a war on consciousness itself. The "war on drugs" is a strange decoy label.' The last bit might not resonate with most people of a scientific bent, but any nerd should have sympathy for what Hardison says is the simplest reason for his actions: 'My love of learning.' Crucially, Hardison said his initial reason for making and spreading the drug were the same as Wilson's: overcoming the ravages of alcoholism.

But anecdotes are just anecdotes. What does data say? Psychiatrists in the 1950s carried out structured clinical trials with alcoholics and LSD. They wondered if the molecule might spark the personality changes that would help boozers give up the bottle. British psychiatrist Humphry Osmond set up an entire centre in Saskatchewan, Canada, explicitly for this purpose.

Many of these studies verified anecdotal reports: a psychedelic experience did help many alcoholics to quit entirely, or at least scale back their drinking to the point that it was no longer problematic. But these were small trials, a few dozen participants each. Nothing like them has occurred since, due to the clampdown on therapeutic research with LSD. So the idea was left to gather dust.

Neuroscientist Teri Krebs and her colleague Pål-Ørjan Johansen of the Norwegian University of Science and Technology in Trondheim gathered every old study they could find and did what public health statisticians call a 'meta-analysis': a study of the studies.[89] This is considered the most powerful form of statistical scrutinisation possible. With almost any drug and any condition, studies generate variable and conflicting results because people are complicated, everybody is different, and a huge variety of factors – from diet to upbringing, the weather, and what you saw on television that day – influence your health. Especially your mental health, complex beast that it is. Collecting all the studies, pooling the numbers and obtaining an overview is a powerful probe.

Johansen and Krebs crunched the numbers, and came up with this: of 536 people in six trials, 59 per cent of people who received one dose of LSD reported lower levels of alcohol intake, while 38 per cent of those who received a placebo cut down on drinking.

'We were surprised at how consistent and lasting the effect of LSD on alcoholism was across trials,' says Dr Johansen.

Alcoholism is just one of a number of conditions notoriously resistant to conventional treatments which psychedelic intervention may help to ameliorate. In the same vein, Dr Roland Griffiths in Baltimore is looking at using psilocybin-containing mushrooms to help smokers quit.[90]

Of all potential therapeutic applications for psychedelics, one has the most promise: treating post-traumatic stress disorder (PTSD) with MDMA. PTSD can be devastating, and 20 per cent of sufferers do not respond to any other form of treatment.

The first legitimate study to examine the use of MDMA combined with psychotherapy for sufferers of PTSD commenced in 2010[91] and more studies are published every year.[92] Scientists are keen to stress: MDMA itself is not a silver bullet for PTSD, the key is to administer the drug in a psychotherapeutic setting to help patients confront difficult memories.

The idea is not new. Alexander Shulgin introduced psychotherapist Leo Zeff to the molecule in 1977, and Zeff in turn introduced thousands of therapists to the drug.

Psychiatrist Dr George Greer[93] was one of them. He conducted some of the first trials with the drug, administering MDMA to more than 80 people in the early 1980s before the drug was banned. 'At this time, the supply of MDMA was in compliance with health and drug laws, and we were keen to keep it hush-hush after what had happened with LSD in the 1960s,' he recalls. 'But you can't keep something like this secret forever. When I heard in 1982 that people were having big parties with it I decided to get knowledge out into the scientific literature.'[94]

Why does it seem to work so well? 'There are no conventional drugs that put somebody in an altered state of mind to facilitate conventional psychotherapy. The point is to use this as a single-shot facilitator of a psychotherapeutic treatment, rather than as a drug to change your mood on a daily basis,' says Dr Greer.

MDMA is soon to be used in clinical trials investigating the use of this chemical 'empathogen' for autism. This occurred as a result of the widespread availability of the drug outside the preserve of the medical establishment, a case of what Rick Doblin calls 'crowd-sourced drug development'. People on the autistic spectrum began to swap stories online describing how the compound helped their social navigation. They collectively campaigned for mainstream medical studies.

Similarly, victims of cluster headaches demanded the same after anecdotally discovering that LSD could alleviate their crushing pain, said to be so unbearable that they obliterate the desire to live, lending them the nickname 'suicidal headaches'. And their campaign was successful: acid is now being tested as a treatment.[95]

This points to a crucial issue: who should have the power to decide if you can take a drug?

Your government?

Your doctor?

Your friends?

Your intuition?

The pharmaceutical companies that produce purified pills and market them to you with soothingly reassuring advertisements?

The renegade chemist next door who cooks quirky contaminated chemicals and peddles them to you with seductively convincing proclamations?

How can you know who to trust? And how can you decide when you are desperate for psychological relief? Hopelessness breeds vulnerability.

In the face of overwhelming despair, many have opted for treatments far more devastating than crack cocaine.

## SURGERY OF THE SOUL

Psilocybin-containing mushrooms seem to prove useful in the treatment of obsessive–compulsive disorder[96] – the only condition left which neurosurgeons will still consider treating with a prefrontal lobotomy.

Today the word 'lobotomy' is synonymous with 'butchery'. But we frequently forget that this operation reaped devisor Dr Egas Moniz the Nobel Prize in physiology or medicine in 1949. American physician Dr Walter Freeman (1985–1972) attempted to make the treatment quicker and safer by inserting medical tools into the brain by going under the eyelid, rather than boring through the skull. Instrument of choice: an ice pick. Without gloves.

In the 1950s, the unhappy and unwell lined up for this 'surgery of the soul', paying neurosurgeons to swirl ice picks through their brains to sever connections between the front of the cortex (considered to be the source of higher thought and complex thinking) and the 'animalistic' regions of the brain further back. It was thought that impingement of ancient 'emotional' areas of the

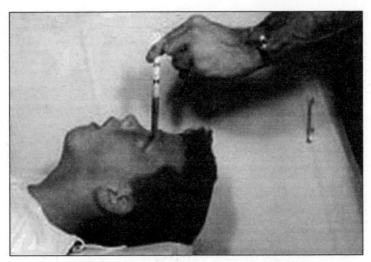

41. More than 50,000 people underwent transorbital lobotomy – delivered via icepick inserted under the eyelid, rather than a tool bored through the top of the skull – in the middle of the twentieth century. Howard Dully, just 12 when he was subjected to the operation in 1960, is one of the youngest known living cases.

brain on the more 'rational' parts of the brain led to many forms of mental illness, from anxiety to schizophrenia and postpartum depression.

Today the procedure is considered appropriate for only one ailment: obsessive–compulsive disorder. Sufferers at the extreme end of the scale will scrub, pick and claw at themselves with such ferocity that they shred flesh to the point that bone becomes visible. Recent studies indicate psilocybin-containing mushrooms could be useful as an alternative intervention. This may be unsurprising for anyone who has enjoyed an afternoon with the blue-stemmed fungi. The giggles, elevated mood, and sense of calm are exactly what made the fruiting bodies of *Psilocybe semi-lanceata* so popular in the UK until the government reclassified them as Class A narcotics in 2005.

Researchers are also investigating ketamine as a means to heal the mind, which is remarkable considering how loathsome the chemical is to those who dislike it. 'It's like the marmite of drugs,' laughs psychologist Dr Celia Morgan[97] of UCL, who has researched ketamine and the psyche for 15 years.

When she first began to investigate the chemical it was as an 'emergent drug of abuse'. It was new on the scene. Now research shows it could work as an antidepressant. 'It has been very enlightening to see this turn into a potential treatment,' says Dr Morgan. 'Ketamine users were initially turned away from clinics and told flatly that it was not an "addictive substance".'

Contrary to what authorities once maintained, ketamine is extremely addictive, making it rather 'weird' among psychedelics, says Dr Morgan. The users she works with consume up to five grams daily. 'Often because they self-medicate for depression or they got into drugs in a psychonautic "exploration of the self" and became trapped in a cycle of repeated use.'

One of the earliest examples of ketamine addiction can be seen in the scientist and psychedelic theorist Dr John Lilly (1915–2001), who 'completely destroyed' himself with ketamine, as his colleague Rick Doblin puts it. As he became more and more dependent on the anaesthetic, his theories became weirder and weirder. Using isolation tanks to communicate with dolphins for example.

The first study to hint that ketamine could be useful in ameliorating depression was published in 2000 in the journal *Biological Psychiatry*.[98] Another published in 2006 in the *Archives of General Psychiatry*[99] came to the same conclusion. Double-blind tests with 17 acutely depressed patients who had not responded to conventional antidepressants were given either ketamine or saline solutions. Recipients of ketamine found their depressive symptoms improved after just 72 hours. This last point is crucial:

medications like Zoloft or Paxil take weeks to kick in (if they work at all). Ketamine trickled into the deep circuits of subjects within just three days.

The other important point is the role neurotransmitters play in depression. Most people are familiar with serotonin and the primal role it plays in mood. Ketamine does not act on the serotonin receptors, like the psychedelics psilocybin or LSD-25, nor is it mediated by the GABA networks, like alcohol or barbiturates. Instead, it preys on receptors that normally receive the neurotransmitter glutamate.[100] This adds weight to the argument that depression comes in a diversity of flavours, attributable to a variety of biochemical triggers. There is unlikely to ever be one pharmaceutical silver bullet.

Animal studies indicate that ketamine triggers 'synaptogenesis' – the formation of new connections between neurons.[101] It seems to be this mechanism that is responsible for the antidepressant effects of the drug, says Dr Morgan. This will be important, because traditional drugs for depression do not work for all. 'We need new treatments – people are still bogged down in SSRIs,' she explains.

Not everyone responded positively to ketamine – so strong it is more commonly administered to horses – but the results were still noteworthy: 64 per cent of depressed participants in one study[102] said they felt better the day after taking the tranquiliser. All were acutely depressed and had not responded to conventional treatments. They took part in experimental trials with dangerous chemicals because they were desperate.

Now, this may come as a surprise to many. No other molecule has eviscerated my generation more than ketamine. The chemical is devastating to the lining of the bladder – the proper diagnosis is termed 'ketamine-induced ulcerative cystitis'. Tooth loss, damage to the urethra, and worse can result.

Yet it can alleviate suicidal thoughts.

This, says Rick Doblin, gets to the heart of what a drug is, when it can help, and when it can harm. 'This was the main mistake of prohibition: to say there are "good drugs" and there are "bad drugs". There is no such thing as a good or bad drug: it's about how you use them.'

Does this mean LSD will soon be widely adopted as a tool to combat alcoholism? MDMA for PTSD? Psilocybin for OCD?

Unlikely.

'LSD and other psychedelics are not patentable and therefore not candidates for daily medication. So no pharmaceutical company will back the research,' says Dr Johansen.[103]

His is a common stance among scientists who research psychedelics: the biggest obstacle to their reformation from black market treats into regulated medications is the pharmaceutical industry's reluctance to fund their investigation, manufacture and distribution. Most psychedelics are out of patent (MDMA, for example, being first fabricated in 1912), or come as a single-shot dose, rather than a daily pill, and thus not likely to be profitable.

Dr Johansen is quick to point out that the clampdown on psychedelic research by authorities worldwide was not as thorough as most people think.

'LSD has never been *banned* outright from medicine. Sandoz simply never applied for marketing approval. When the patents expired in 1965, Sandoz no longer had an economic incentive to manufacture LSD-25,' he explains. 'The National Institute of Mental Health in the US and the National Institutes of Health in the US actually *rescued* LSD research. They provided the LSD-25 samples for our clinical trials.'

# SHAKING UP A SNOW GLOBE

The resurgent scientific usage of psychedelics in neuroscience could have a higher application: understanding how the brain works. Hedonistic indulgences are leading us to profound intellectual insights concerning the mechanics of the most complex thing in the universe.

Just as nicotine led us to discover the keyholes for the molecular messengers of the brain, caffeine aided our decoding of the chemical commerce of the cell, marijuana and opium both led to the discovery of our body's inborn painkillers, hallucinogens today are proving to be what LSD's discoverer Albert Hoffman hoped they could be: telescopes for the mind. Our predilection for what may seem base or indolent pursuits are once again catalysing sophisticated scientific discoveries.

'You can only really learn about any phenomenon by probing it – taking a psychedelic is a bit like shaking up a snow globe,' says Dr Robin Carhart-Harris of Imperial College London. In 2012 he published a landmark study, measuring the flow of blood through the brains of people dosed with psilocybin using the magnets in an fMRI scanner.[104] Contrary to what many might have anticipated, Dr Carhart-Harris discovered that the hallucinogenic psilocybin doesn't increase the flow of blood to the brain. It *reduces* it.

Carhart-Harris's studies have indicated that the hallucinations, sensations and happiness magic mushrooms can produce are caused by decreases in blood flow through what are called 'the maximal hub regions', including the thalamus, the 'anterior and the posterior cingulate cortex,' and 'the medial prefrontal cortex (mPFC)'.

'A lot of the ways our brain operates during normal waking

42. The hallucinogens LSD and psilocybin both work by fitting into our body's receptors for serotonin – one of the most famous neurotransmitters in existence – better than serotonin itself.

consciousness seems to be aimed towards keeping our experience of the world to a precise minimum,' Dr Carhart-Harris explains. 'If psychedelics reduce activity in important regions of the brain, this suggests that activity in those regions keep our experience of the world very stable. Normal waking consciousness is actually an optimal mode of consciousness. The constraining nature means some things are lost. Everything is a trade-off.'

Executing this study was not easy. Psilocybin, though it grows freely in the fields of England, had to be obtained from laboratories in Switzerland at exorbitant costs. Licences, red tape and regulations are a nightmare. 'The ultimate effect of all regulations is to harm science,' he says.

Brazilian neuroscientist Dr Draulio de Araujo's research, administering ayahuasca cocktails to healthy human subjects and sticking them into brain scanners, is far easier and far cheaper thanks to the legal use of the brew in Brazil. His work shows that the hallucinogenic cocktail has similar effect to psilocybin, decreasing blood flow in the same constellation of brain regions. He and Dr Carhart-Harris call this the 'default mode network', speculated by some to be 'the seat of the self'.

There was great hope in the 1950s that psychedelics would unveil the mechanics of the brain. As Hoffman put it, psychedelics could be for psychiatry what the telescope was for astronomy. The middle of the twentieth century saw an important illustration of this idea: discovering the role of serotonin in the brain.

The neurotransmitter serotonin had been known to scientists since the 1860s, as it is found in high concentrations in clotted blood. Its chemical structure was elucidated in 1949. Biochemists Maurice Rapport and Arda Green personally lugged two tons of coagulated beef blood daily from a Cleveland slaughterhouse – bucket by bucket – to provide the raw fodder from which the compound could be extracted.

Though serotonin is today a celebrity among the messengers of the mind (selective serotonin reuptake inhibitors, SSRI antidepressants, work by blocking the reception of this chemical cargo), it was not initially thought to be important in the functioning of the brain. Roughly 90 per cent of the serotonin in the human body is found in the intestinal tract. Thus until the 1950s, biologists thought serotonin was merely a chemical cast member in the gut. The discovery that serotonin also works in the brain was made by two teams, one American and the other Scottish, in 1953.[105]

Young and female, Betty Twarog's (1927–2013) proposition that serotonin might be found in the brain was predictably greeted with much grunting by her male Harvard overseers. The game changer came when biochemists noted that serotonin, aka 5-HT, looked an awful lot like the new compound that had been causing such a fuss among their psychiatrist colleagues: LSD-25.

If LSD catalyses profound transformations in the mind, and serotonin has the same chemical backbone as LSD, could serotonin therefore play an important role in the brain? Pharmacologists – Britain's Sir John Gaddum and D. W. Woolley and E. Shaw in New York – all arrived at the same conclusion: the action of LSD seems to be mediated by the same chemical networks that transmit the signals of the neurotransmitter serotonin.

The subsequent investigations into the role of serotonin in the brain proved to be one of the most productive streams of research in the history of neuroscience, leading to the development of antidepressants that save lives, cause controversy, and fill our pharmacies.

But this is one of the few scientific victories for psychedelics: research into the mechanisms of the mind using these banned substances slowed to the pace of an LSD-25 dosed snail after 1970.

The hope is that now, with the so-called 'psychedelic renaissance' taking place, neuroscientists will be able to shake and probe the brain with new chemical and technological tools. With the advances in brain-imaging technology that have taken place since the silence of psychedelic science, who knows what they will find.

One thing they will, without a doubt, further elucidate: our brains shape the world we live in. We do not see or feel the world as it is – we experience our brain's computation of it. And few things prove this more than the capacity of drugs to alter our perceptions.

'Psychedelics remind you that you have a brain at all,' as Dr Carhart-Harris puts it.

# BETTER DYING THROUGH CHEMISTRY

One of Hoffman's greatest hopes for LSD-25 was to 'transform the experience of dying'. His aspiration transformed the expiration of a select few. Aldous Huxley famously called for the molecule to be brought to him on his deathbed.

We may not yet have reached the point that lysergic acid will be administered in hospices or hospitals, but American clinicians have scrutinised psilocybin for its capacity to treat anxiety and depression in terminal cancer patients.[106] 'The overwhelming demoralisation in these patients is extremely hard to describe, and even more difficult to treat,' says Professor Charles Grob of the Department of Psychiatry and Biobehavioral Sciences at UCLA.[107]

Other researchers are using psilocybin to investigate emotional experiences of another sort: religious revelations. A 2006 study at the Johns Hopkins University School of Medicine in Baltimore administered psilocybin to 'spiritually inclined' volunteers who had never partaken in psychedelic use. Fourteen months later, 67 per cent of the participants rated the treatment as one of the five most profound experiences of their lives.[108]

No compound has been subject to more spiritual speculation than DMT, which we met earlier in this chapter lacing the cigarettes of monkeys locked in dark boxes. Some studies indicate DMT is an 'endogenous' hallucinogen: unlike other imported narcotics, it may be a native inhabitant of our bodies.[109]

Other drugs hijack the docking bays of our communication networks. Cannabis disguises itself as our endogenous endocannabinoids. Opiates siege the gates of our endorphin circuits. Caffeine usurps the energetic pathways of our nervous system. Psilocybin and LSD-25 both spookily bind to our serotonin receptors more strongly than serotonin itself in an impressive act of hallucinogenic insurrection.

But DMT seems to be unique as a home-grown biochemical inhabitant. Not a psychedelic intruder: a neurochemical native. Psychopharmacologist Dr Rick Strassman is a key figure in the history of legitimate peer-reviewed scientific research into hallucinogens. He instigated the first human studies with DMT in the early 1990s, when few scientists would dare experiment with psychedelics on humans. Strassman's work indicates that not only is DMT a native resident of the body, but is produced by the unrivalled pineal gland.

This tiny singular nugget has been scrutinised by philosophers and anatomists for centuries because it is the only 'unpaired' region of the brain: unlike the hippocampus, amygdala, nucleus accumbens or sensorimotor cortex, it exists as a singular entity: there are no left and right versions. There is only one pineal gland.

The pineal's scrutinisation has a long and illustrious history: Descartes called it the 'seat of the soul'. More recent neurological explorations have revealed that, in a sense, the pineal is not a part of the brain at all. It develops in the roof of the fetal mouth then migrates into the brain, nestling just below the pituitary gland, the mischief-maker of puberty. It's an anatomical rogue.

Dr Strassman's meticulous records of his patients' descriptions of their adventures on DMT[110] – ranging from heaven and hell to alien spaceships – makes for lively reading. (He always measured their body temperature with rectal thermometers. One wonders if this influenced the nature of psychedelic visions.)

'The pineal gland could act as an antenna or lightning rod for the soul,' Dr Strassman writes. 'Psychedelics affect every aspect of our consciousness. It is this unique consciousness that separates our species from all others, and that gives us access to what we consider the divine above. Maybe that's why psychedelics are so frightening and so inspiring: they bend and stretch the basic pillars – the structure and defining characteristics – of human identity.'[111]

Strassman has devoted his career not just to studying DMT but also exalting it: he asserts that the action of DMT through the pineal gland unifies all spiritual experiences. He has written an entire book celebrating the idea, *The Spirit Molecule* (2001). A film of the same name catalogues reports of adventures to the centre of the universe and cavorts with little green men.

As noted, descriptions of other people's drug experiences are almost invariably boring. Why this should be is counter-intuitive. Tales of escapades with 'machine elves' scuttling through the bowels of intergalactic spaceships, out-of-body adventures spent watching oneself from five metres to the left, and sexual encounters with celestial bodies all ought to make for stories that are anything *but* boring.

But listening to people describe their drug experiences tends to be tiresome. One does not have to be a veteran of late night fireside hippy chats to grow weary of stories that begin with the ingestion of a chemical and end with a grandiose conclusion: 'What if DMT is the key to understanding LIFE on other PLANETS?'[112] Yawn.

But I have one story from my own misadventures that I feel is worth sharing.

Sitting on a beach in Devon in August 2003, I contemplated grumpily in the sparkling reflection of the swirling ocean that I would return home in three days to Toronto. University, a relentless bar gig, newspaper job, and the usual routine of five hours sleep a night, massive endocrinology textbooks and painfully dry statistics modules. And the upcoming, demoralising Canadian winter. I wasn't excited.

'I think I have the solution to the way you feel,' said a friend. 'I've got some DMT: it's pretty much the world's strongest hallucinogen.'

After pondering, I agreed to try it. Why not? Just this once.

I followed his instructions, smoked three tiny grains from a pipe, and fell to the ground in a retching fit, vomiting profusely into the sweet English grass. I emptied the contents of my stomach unceasingly, internally cursing myself for my gullibility. (It felt like hours, but my friends all concurred: the wretching lasted no more than 20 seconds. Such is the nature of psychedelic delusion.)

Once the sickness subsided, I lay back with my eyes closed, gasping for breath, and tried to breathe deeply. Opening my eyes, I looked down at my prostrate body. Every inch of me was covered in grasshoppers. They waved their antennae at me casually, and looked at me with bright compound eyes.

I looked behind me at my friends, and though I had been rendered non-verbal, the look in my eyes clearly communicated what was on my mind ... really? Surely this was the spectacular illusion of one of the world's strongest drugs.

'You are NOT hallucinating,' said one friend. 'You are *actually* covered in grasshoppers.'

'They love you!' another enthused.

Is Dr Strassman right? Does DMT wind through all living things – from grass to grasshoppers, hippies, hippos and humans – and is it the key to all spiritual experience?

There is no hard evidence to back up Strassman's beliefs that DMT has some kind of magical property, binding all living things together – and I'm dubious about his claims.

All I can say is that I smoked three grains of his favourite chemical, and found myself covered with grasshoppers. Needless to say, I don't have a peer-reviewed study to validate the experience with statistical significance. But I do have four friends who saw it with their own sober eyes.

I have been offered DMT many times since – and I have never taken it. Once was enough.

Most stories about other people's drug experiences are boring. I hope this one wasn't.

# ROCK 'N' ROLL

# INTOXICATING INFLUENCES

At the interface between drugs and music, scientific investigations have been enlightening, most particularly in the records of neuroscience. Music and drugs both tickle many of the same neural networks, most importantly, a tiny organ called the 'nucleus accumbens'. They both result in the release of many of the same neurotransmitters, including our old friends serotonin, dopamine, oxytocin, and a host of endorphins. We'll come to how scientists dissect the innards of the brain – or 'peek under the hood', as David Byrne describes it[1] – later.

Legitimate scientific explorations of the impacts drugs have on musical style and compositional form are rare, for two reasons. First, understanding the link between substances such as ketamine or MDMA and the tempo of dance music does not tick as many boxes for eagle-eyed funding application judges as the potential medical use of such molecules. And second, the links are obvious if one examines the chronology of musical evolution. Independent empirical investigations will not be useful when history already provides an instructive repertoire that illuminates the influence of specific compounds on particular musical styles.

But this does not mean that scientists are unaware of the link between the two.

To address the 'lack of engagement between researchers, healthcare professionals, users and policymakers',[2] scientists from University College London sought to bring regular ketamine users into the same arena as medical academics through a day of music, film and art dubbed – appropriately – K Day. They should have called it 'K Pride'.

Ketamine addicts at the time were rare in London. So UCL had to bus in a number of willing 'krusties' from Bristol with the same funds used to pay for other ostensibly more legitimate research programmes.

Psychologist Dr Celia Morgan of UCL had an astute aim: enlisting participants in clinical studies to research the long-term impacts of heavy ketamine use, under-researched and increasingly devastating.

On offer: a range of stalls titled 'Ketamine and Your Brain', 'Ketamine Cystitis', 'Pain and K-Cramps', 'Release, Drug Use and Law (Free Legal Advice)', 'Nutrition', 'Detox', 'Psychology and Psychiatry in Drug Services', and 'Respect Drug Users Group' administered by recovered addicts. Academic researchers also constructed 'Kreativity Korner', where ketamine users could explore 'art and music therapy to creatively describe their experiences'.

'It did get a bit out of hand,' laughs Dr Morgan in recollection.[3] 'Let's just say this: the line-up for the toilets was very long.'*

K Day was held in Covent Garden, once the flower stall centre of London. Americans will know it from *My Fair Lady* (1964), today a middle-class shopping centre purveying lattes, merino wool scarves and gluten-free cupcakes. Except for one day, when it housed Kreativity Korner and dreadlocked drug users lined up to snort powdered horse tranquilisers in the toilets.

The mess may have been a headache, but the event did succeed in delivering life-saving information to heavy users who genuinely needed it, including advice on mindfulness and other tools former addicts have used to kick the junk. For her part, Dr Morgan did succeed in gaining the trust of heavy ket-heads, who subsequently signed up to participate in formal research.

'Also: I bet the music was good,' I said.

'Oh yes, it was,' she laughed.

---

* This means that they were pouring piles of powder on to the toilet lids so they could surreptitiously snort lines.

# SUBSTANCE & CHARACTER

We all know drugs influence the style of music, yet legitimate scientific investigations into the impact drugs have on musical style are few and far between. History is the only reliable record we can scour for evidence that illuminates how drugs change the composition of music.

Every generation likes to think that it invented getting high. One of the nineteenth-century's most colourful characters – literally colourful, owing to the hue of his red nose – was composer Modest Petrovich Mussorgsky (1839–1881), one of the main drivers behind the evolution of romantic music. He wove together overblown orchestral crescendos long before Peter Ilyich Tchaikovsky (1840–1893) or any of the better-known composers who followed in his footsteps. He was an angry alcoholic (hence the nasal colouration), or, as his ailment was termed by the burgeoning field of psychiatry, 'dipsomania'.

The boozy nature of his temper is loudly reflected in the intensity of his work, such as his best-known piece, 'Night on Bald Mountain'. You might know it from the nightmare sequence depicting Satan conjuring an army of spirits in Disney's *Fantasia* (1940).

Jazz musicians were notoriously enamoured with opiates. As the saying goes, 'ain't got no funk if you ain't got no junk'. Notably, Miles Davis, notorious addict, was apparently always sober in the studio. The pace of punk noticeably speeds when amphetamines became the scene's drug of choice, especially in the UK. Earlier bands such as the Stooges and the New York Dolls played at slower tempos and were more into heroin. Stoner rock is made with exactly what it says on the tin: cannabis.

Reggae is written with the resin of marijuana, and the more cannabis consumed, the slower the pace, as evidenced by the dramatic change in the metre of Bob Marley & The Wailers when Lee Perry joined. Perry has said he would never have formulated dub at all were it not for ganja.

Cocaine's influence tends to be energetic but unoriginal. Disco is characterised by clean chords and polished production, omitting the sonic experiments of their psychedelic forerunners. Sometimes the outcomes are, if not interesting, admittedly impressive. Fleetwood Mac's 1979 song 'Tusk' is a good example: it was recorded live in an American football stadium – exploiting the acoustic qualities of the venue – with the University of Southern California Marching Band.

Few genres wear the chemical stain of inspiration on its sleeve quite like electronica. The roots of slow-paced dubstep were sown with ketamine, an anaesthetic and tranquiliser. Long before this, house music was born in the warehouses of New York through the injection of ecstasy pills into the gay club scene. The combination of MDMA with amphetamines, a hyper but happy duo, reliably leads to repetitive 4/4 beats that sober people often find monotonous. For this reason, avant-garde modern classical composer Steve Reich, who profoundly influenced dance music, is presumed to have dabbled (despite all his public denials), because of the phenomenally repetitive nature of his compositions.* Monotonous 'Music for 18 Musicians' may be, but it brilliantly exploits the brain's habit of imposing patterns on seemingly random information.

Throughout the 1980s and early 1990s, England saw itself host to gargantuan illegal raves, largely housed in the abandoned industrial warehouses of the north. Conservative Prime Minister John Major attempted to prohibit such gatherings through the Criminal Justice and Public Order Act of 1994, which sought to ban assemblies paired with 'repetitive beats', is endearingly naive.

---

* http://www.youtube.com/watch?v=xU23LqQ6LY4

Few drugs however had as profound an influence on the musicians who took them than psychedelics – and they continue to do so. There's the most obvious example: the story of the clean-cut kids from Liverpool who make it big with clean cut conventional rockabilly, then are introduced to LSD in 1965. Then they don bright red jackets and sing 'I am he as you are he as you are me and we are all together' in 'I Am The Walrus' (1967).

More extreme examples abound. Captain Beefheart (1941–2010) once drove his entire band into the desert while he was on LSD-25 and left them there so they could understand 'the vibe he was going for'. There is some speculation that he suffered from schizophrenia, which may be true, but what is without question is that the elephantine volumes of psychedelics he habitually consumed would not have aided his mental balance. His drug use was not publicly broadcast at the time. He told *Rolling Stone* magazine in 1970 that neither he nor the band *ever* took drugs.

Others were more overt. The Grateful Dead made a point of playing live on LSD. Richard D. James, also known as Aphex Twin, titled his 2001 album *drukQs*. The inspiration for The Flaming Lips 1989 demos *The Mushroom Tapes* is fairly obvious. American rapper and techno DJ Flying Lotus (grandnephew of Alice and John Coltrane) expounds on the virtues of DMT in printed interviews as a matter of course, and during live sets implores the audience to chant 'DMT'!

Spiritualised get the gold medal for actively encouraging the consumption of drugs, describing their work as 'music to take drugs to', and their creative process as 'taking drugs to make music to take drugs to'. Their seminal album *Ladies & Gentlemen We Are Floating in Space* (1997) even came wrapped in mock prescription drug blister packs, just in case anyone hadn't quite got the message.

No discussion of prog rock or psychedelics would be complete without Pink Floyd. And Syd Barrett. The original lead singer for Pink Floyd, Syd left the band in 1968 after the release of their first album and their stratospheric launch into fame due to mental health issues exacerbated by drug use. By 1970 Syd was a recluse. He remained a hermit until his death in 2006. Barrett was one of the world's first well-known 'acid casualties'. On the heels of the Beatles' popularisation of psychedelics via flowers and rainbows came Barrett's transition from an experimental, groovy, gorgeous London swinger into a long-haired, wild-eyed, gaunt, sad shade of his former self. A sobering reminder that every single drug is a double-edged sword.

The rest of Barrett's life is a sad story. The exact nature of his mental illness – schizophrenia, bipolar disorder, some form of autism – is unknown, but it's widely accepted that psychedelic straws broke the camel's back. Syd lived out the long, lonely years of his life after his excess. Many others didn't. Society undoubtedly holds musicians who die in the throes of drug abuse in a state of reverence. Their demise confers a divine form of canonisation. But would we still think of them so fondly were they alive today? What if Jimi Hendrix had gone broke and wound up licensing songs to Coke?

## REDEMPTION AND RECOVERY

Not everyone, of course, suffered an early demise or drifted into narcotic obscurity. John Coltrane's opus *A Love Supreme* (1965) was inspired by a near overdose, but he survived and is now remembered for his music, not his death. The Velvet Underground's Lou Reed (1942–2013) once penned a song simply titled

43. Cyber metal legends Gwar behead puppets and cover the audience with fake blood. When my father, a promoter, staged them he received complaints from angry parents for two weeks about the red stains on their children's contact lenses. True story.

'Heroin' (1974), but he kicked the habit. Former junkie Iggy Pop is still ticking, still touring, and is now advertising car insurance.

David Bowie's 'Space Oddity' (1969) – which represents a hit of heroin, not a journey into near orbit – was followed by a decade of success and excess. A retreat to Berlin to sober up resulted in the album *Low* (1977) during his recovery, followed by *Heroes* that same year. The title track explodes from every stanza with the life-affirming sense of triumph that only reformed addicts have.

Drugs are not required for the creation of good music. That much is obvious. Aretha Franklin never needed crack or LSD to render her singing sublime. Same goes for Otis Redding. Ditto for the equally sublime Sam Cooke. There are thousands more.

But perhaps the real question is: are drugs crucial for musical *experimentation*? History tells us that they can help. But are they necessary?

Stevie Wonder's unparalleled career has seen him take many forms, from a clean-cut harmonica-playing tot to a beaded-hair hippy, and he never needed drugs for inspiration. He says he only smoked marijuana once – and hated it. This is noteworthy considering *Innervisions* (1973) is profoundly psychedelic.

Cyber metal originators Gwar, who dress as warrior lizards on stage, behead giant puppets, and cover the audience with fake blood (seriously, they're awesome), are comic-book geeks backstage who subsist mostly on Doritos and Coke. This is true of many of death metal's most extreme acts.

Tom Waits is an interesting example: he issued beautiful but fairly conventional jazz on his first two albums, *Closing Time* (1973) and *The Heart of Saturday Night* (1974). Much alcohol was imbibed, especially during the following years on tour. But then he went teetotal – and ever since has conjured strange carnival-inspired odes to the world of circus freaks. With every album, he just gets weirder. 'There's nothing romantic about being a drunk,' he has said.

'The idea that creative endeavours and mind-altering substances are entwined is one of the great pop-intellectual myths of our time,' Stephen King writes in his autobiography *On Writing* (2000), recounting in every gory detail his own battles with alcohol and drugs. His description of his children pouring a pile of 'blood-spattered coke spoons' in front of him for his scrutiny is forever scarred in my mind.

Perhaps the most telling example can be found in experimentalist Frank Zappa. His music is notoriously weird and his mind was unremittingly sober. Incidentally, he was a close friend of Captain Beefheart, an unparelleled wreck-head and published many of Beefheart's albums on his label Straight Records.

Frank Zappa never took drugs and he named his child Moon Unit. That says it all.

## TECHNOLOGICAL TRANSFORMATIONS

In 1893 inventor Thaddeus Cahill, devisor of mechanisms for typewriters and pianos, became obsessed with a grandiose idea.[4] Might it be possible to combine engineering innovations with developments in the scientific understanding of electricity to produce music? And send it to another location? A few tinkerers before him had tried, but all previous attempts to broadcast the sounds of live musicians through telephone communication networks had failed dismally: the sound was too weak, too tainted with white noise and static.

Might it be possible to create music using electricity alone?

Thus the world's very first electronic instrument was born: the Telharmonium (also known as the Dynamophone), an electronic organ, constructed two decades before radio brought music to the masses. The Mark 1 Telharmonium, completed in 1896, weighed seven tonnes. It's successor came in at a whopping 200 tonnes.

Electrical amplification was still unknown at this time, and vacuum tubes had not yet hit the scene. Cahill came up with the idea of building complex tones from a series of electrical generators. Scottish engineer Lord Kelvin (1824–1907) was among the lucky group of industrialists and bankers treated to the instrument's debut performance. 'One of the greatest accomplishments of the brain of man,' he declared.[5]

The second model, the Mark II, was financed with more than $100,000 from financial backers, most notably Oscar T. Crosby,

44. The world's first electronic instrument, the Telharmonium.
The seven tonne organ, unrivalled in its construction, was
created 20 years before the radio. Alas: it sounded terrible.

who had funded the development of America's telephone networks.[6] He and select entrepreneurs, bankers and engineers were
treated to an exclusive presentation of the instrument: they heard
it broadcast from 35 miles away to demonstrate the organ's unique
capacity to transmit music remotely. They were bowled over.

The 200-tonne Mark II's mainframe spanned 60 feet with ten
switchboard panels, covered in more than 200 switches. Each note
was produced by a complex set of electrical generators, together
capable of producing up to 14,000 watts of power. Two musicians
were required to play it – and when the motor slowed, power the
belt manually. Massively heavy, the instrument required a fleet of
ships to transport it from the site of construction in Baltimore to
a theatre on the Rialto in New York City.

The 60-foot long banks of generators were housed in a base-
ment beneath the keyboard, where a brambling nest of wires

45. Today an mp3 player can carry tens of thousands of songs on a piece of metal that can fit in your pocket. In 1893, a generator the size of a child was required to produce one note.

connected the mainframe to the dense garden of rotors. Each was four feet long.

Cahill believed he was creating 'pure and clean' tones that could replicate the frequency patterns that demarcate 'real' musical instruments from each other. Eighty years before electronic synthesisers did the same, he designed the Telharmonium to mimic not only the piano, but also the flute, bassoon, clarinet and cello (albeit none of them terribly convincing impressions). Like today's digital fetishists, he thought electrical purity could replace all antique forms of musical production.

But Cahill and Crosby were not really interested in making beautiful music: they hoped to make a fortune by replacing restaurant musicians, who at the time were costing the bistros of American cities a fortune, up to $1 million a year. In competition for clientele, New York's restaurateurs were forced to employ up to 40 musicians apiece to play in private orchestras, and their wages were among the highest in the industry. Cahill and Crosby thought they could market the Telharmonium as a cheap alternative, replacing live ensembles and their salaries: the electronic music equivalent of the spinning jenny. Just two people playing the instrument in the Rialto could simultaneously broadcast music to hundreds of cafes, dining halls and theatres.

On paper, it sounded great. The instrument was in every way unlike anything anyone had ever seen. There was just one problem: it sounded terrible, like a street organ crossed with a malfunctioning synthesiser. Technical hurdles also cropped up. Crosstalk between networks resulted in telephone users being subjected to strange noises bleeping into their conversations.

Eventually the massive size of the instrument proved too insurmountable a burden for operators to maintain. The Telharmonium lived for just two decades in New York before the company was forced to declare bankruptcy in 1914. The original seven-tonne Mark I was finally scrapped in 1962.

The Telharmonium is one of music history's more colourful stories – but it is hardly an isolated tale. We have used every single material available to us – wood and bone, intestine and horsetail, metal, glass, and plastic – to create new means of making music. We have devised new ways to make noise with whatever we could get our hands on from the very beginning.

## ROCK, BONE AND WATER

Some of the oldest known archaeological artefacts are musical instruments, including what could be a flute, more than 40,000 years old. Known as the 'Divje Babe flute', this fragment of a bear's

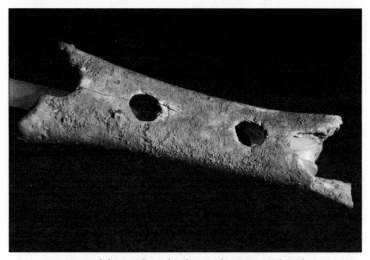

46. Carved from a bear leg bone, the Divje Babe Flute, or 'Neanderthal flute', at 40,000 years old could be the most ancient crafted artefact ever discovered.

leg bone pierced with two equally sized holes was found in a cave in Slovenia in 1995 by archaeologist Ivan Turk. He claimed it was fashioned by Neanderthals, *Homo neanderthalensis,* and thus dubbed it the 'Neanderthal flute'.[7]

But other scholars of ancient hominids counter that it was created by our own supposedly more intelligent ancestors. Slovenian scientist Mitja Brodar believes the instrument was fashioned by Cro-Magnons,[8] the prehistoric predecessors of *Homo sapiens.*

Consider for a moment how a flute works: it is a very complex instrument compared to, say, a bell or a drum. Waves of sound need to bounce back and forth between the ends of the chamber in just the right patterns to produce a musical note. You can easily try to blow into a flute and create nothing but silence, or a dissonant whooshing sonic discharge.

Exactly which early strain of human had the wherewithal to create such a sophisticated instrument strikes at an important question. Were Neanderthals the bumbling brutes we have always presumed them to be? Did they vanish because they were out-smarted by our own supposedly wittier ancestors? Or is it possible our ginger-haired relatives[9] had artistic talents and skills surpassing those of our own ancestors?

Archaeologist and author Professor Steven Mithen of the University of Reading argues that our Neanderthal cousins were more musical than our own human ancestors in his 2006 book *The Singing Neanderthals*.[10] Just how virtuosic they were, and if they had the capacity for language, leads to an important question: which came first, language or music?

As we noted from the study of homosexuality, the violent and combative aspects of interspecies relations tends to prevail in the popular understanding of evolution. The concept of 'survival of the fittest' – a phrase coined by Herbert Spencer by the way, not Charles Darwin) – has always carried combative connotations. Thus it has always been assumed that humans warred with Neanderthals, and we emerged victorious.

Yet 'fittest' need not imply the strongest, smartest, or most aggressive: it can also describe the most physically attractive and sexually seductive. This fits neatly in line with the slang employed to describe a pretty girl: 'she's fit'.

It is increasingly appreciated that we did not outwit or destroy the Neanderthals: we mated with them. As the hippies would have put it: we made love, not war. An analysis of the Neanderthal genome in 2010 revealed that they contributed considerably to our own genetic makeup.[11] This was accomplished by scrutinising 4 billion nucleotides – the letters of our genetic language, A, C, T, and G – from three ancient individuals using the polymerase chain reaction. Thank you Albert Hoffman, thank you LSD-25, and thank you Nobel Prize winner Kary Mullis.

Just how much our ancestors cavorted with Neanderthals is

unclear (we might never really know), but genetic analysis tells us that all human populations outside Africa are 1 to 4 per cent Neanderthal. Could we have inherited our musical instincts as well as our red hair[12] from our thick-browed, ginger-haired cousins? Just maybe.

Back to the flute: detractors claim it is not an instrument at all, and the holes in the 'Divje Babe flute' are mere teeth marks left by a carnivore scavenging on the remains of a bear.[13] One only has to look at the bone to find the 'teeth mark' hypothesis unconvincing: the holes undoubtedly look as though they were deliberately bored into the bear bone. The space between the two holes is 3.5 centimetres and the diameter of each hole just under a centimeter – configurations that are comparable to modern-day wind instruments.

Musicologist Bob Fink argued in a 1997 essay[14] that the flute is the surviving fragment of an instrument that originally bore six holes, which would have been of the right configuration to produce the 'do re mi fa' scale.

He makes a compelling case. In 2011 Ivan Turk, who discovered the bone in the first place, made a replica of the instrument with musician Ljuben Dimkaroski based on Fink's model[15]. Turk reports their reproduction 'reveals the object as an instrument in the proper sense of the word', with a range of two octaves (extended to three by blowing harder into the mouthpiece, as modern flutists do). Thus, if it is a flute, it would not only be the world's oldest surviving musical instrument, but a sophisticated and versatile one.[16]

If we were making flutes 40,000 years ago, then when did our ancestors begin to fashion percussive instruments with wood, stone, and animal skins?

For that matter, when did we start to sing? Before drums and gongs, we probably just used our vocal chords. Perhaps we

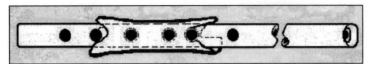

47. Detractors believe the Babje Flute is not a flute at all, claiming
the holes were left by scavenging animals. Musicologists
have shown what the instrument would have looked like
when whole – and what notes it would have made.

called, cooed and hummed before we developed language. As
some anthropologists like to put it, 'we sang before we spoke'.
The change in shape of the human vocal tract marks an interesting milestone in our transition from apes to people.[17] While
chimps and gorillas are only capable of making a narrow range
of noises (albeit very loud ones), humans can produce a phenomenal range of sounds, highly variable in both pitch and texture.
This is crucial for our capacity to both sing and speak.

Over evolutionary time our mouths shrunk, which allowed
for the creation of new sounds. Our necks lengthened and our
larynx (voice box) shifted downwards. With these modified anatomical instruments, we could control the vibration of our laryn-
geal muscles by altering the speed of air pushing out from our
lungs. This evolutionary innovation allows sopranos and tenors
alike to soar to phenomenal ranges and volumes. Otolaryngolo-
gist (ear nose and throat specialist) Alfred Tomatis (1920–2001)
estimated that sound levels within the larynx reach 130 decibels –
equivalent to the sound of a jet engine from 100 feet away. Tomatis's father was an opera singer. And as he observed his father's
hearing deteriorate, Tomatis suspected the warbling could be to
blame. His later scientific work suggested operatic singing – just
like heavy metal screaming today – damages the muscles sup-
porting the bones of the middle ear.

Professor Philip Lieberman of Brown University in the USA

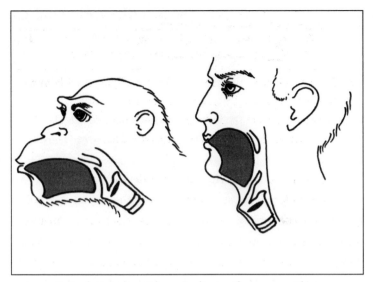

48. Evolutionary evidence indicates that we were born
to sing: the vocal cords (though 'vocal chords' might
be more appropriate) sit low in the throat compared
to other primates, giving us a greater range.

has studied the changes in the human vocal tract for 30 years.
He estimates that our ancestors would have had the anatomical
characteristics to sound just like us 50,000 years ago – possibly
70,000 years ago. (It's difficult to study the evolution of soft tissue
because it does not leave the same indelible mark in the fossil
record as bones do.[18])

Intriguingly, modern acoustic studies on 15,000- to
30,000-year-old cave paintings in Arcy-sur-Cure in Burgundy,
France, have revealed that the highest density of paintings are
located at points within the caves that boast the best acoustics,
resulting in resonance, reverberations and echoes.[19] It is therefore inferred that our ancestors painted cave walls in the exact
locations where they could make the loudest music possible to
accompany their visions.

Even when we were cave dwellers we were sophisticated sound technicians, working with the natural amphitheatres of the world to amplify our tunes. Our ancestors may not have uttered 'CHECK CHECK … MIKE CHECK', but the intention was the same.

Acoustic analysis by French researchers also suggests that ochre markings covering stalagmites springing up from cave floors were carefully placed by our troglodyte progenitors to mark the points which produced different notes on the stone when percussively smacked. Dented rock pillars show signs of having been repeatedly struck, and next to them lie shards of animal bone, suggesting that our ancestors played the cave like a xylophone.[20] Today, experimental musicians (and you yourself if you wish) can play similar 'cave lithophones'.

The most spectacular is the Great Stalacpipe Organ. Built in 1956 in a cave in Virginia by the delightfully named Leland W. Sprinkle, a Pentagon computer programmer, this instrument combines a traditional church organ with a natural geological formation. When the organ's keys are pressed, mechanisms transfer the signal to rubber mallets which play the cave's stony projections like an array of pipes in a church.

One presumes this inspired David Byrne of the Talking Heads creation *Playing the Building*, which has transformed warehouses, railroad roundhouses and brick buildings from Sweden to London and Minneapolis into instruments. Like rocks, rivers and caves, Byrne's buildings prove that anything – anything – can be a musical instrument.

Rock still serves as the foundation for many forms of music today, from the musical stones of Togo, to the gong rocks of Namibia and the 'ringing rocks' of India, North America and Scandinavia. Alternatively, water can serve as a musical substrate. In Zadar, Croatia, the experimental musical instrument

49. We humans can make music with seemingly anything – even the ocean itself. The Morske orgulje ('Sea Organ') in Croatia produces lilting tunes as it surges with saltwater.

Morske orgulje 'sea organ' produces sound through a series of pipes embedded underneath marble escarpments.[21]

The BaAka people of the Congo illustrate that you don't need bone, rocks, caves or pipes to make music. All they need is a river. A combination of slaps on the water's surface, splooshing with their hands and complex patterns of clapping are all the BaAka need to produce a range of distinctive motifs. BaAka 'water drumming' evolved in complete isolation from the rest of the world for thousands of years, so it's little wonder that their percussive rhythms are so complex. And so good.

In every corner of the planet, since the very beginning, we have used anything we can get our hands on to make new noises. Who would have ever thought to extrude the muscular strings from the gut of a cat and rub them against hairs drawn from the tail of a horse? Moreover: who would have ever thought it could sound so sweet? Fiddly and finicky, violins, violas and cellos have changed little for hundreds of years because the strange combination of intestine with keratin remains remarkably beautiful.

50. The BaAka people of the Congo have developed a
form of 'water drumming' in complete isolation from
the rest of the planet. Even without a single piece of skin,
wood or stone in their repertoire, they utterly rock.

Metallurgical bells, wooden resonance chambers, mammalian hair. The global diversity of instruments is vast and colourful. The development of the piano makes for an interesting milestone. Bartolomeo Cristofori (1655–1731), the instrument keeper for the Florentine Medici dynasty, is typically credited as having built the first true piano. But his instrument was modelled on countless experiments in struck strings over past centuries. Traditional stringed instruments such as lyres were plucked by hand, but experimentalists dabbled in designs that divorced the hand from the string. Harpsichords, for example, employed quills for plucking. Numerous experiments led Cristofori to work with what had been determined to be the ideal method: employing a hammer to strike varying lengths of wire.

The pianoforte, which allowed for one person to play up to

ten notes at once, transformed the performance and composition of music. For good reason, it is still one of the most versatile and popular modes of making music: one human can create the same effects as ten. But at the time traditionalists lambasted it as a 'cheap' means to produce melodies. True musical notes, they asserted, were created through the application of human muscle to string. Detractors claimed that allowing a musician to press a button, delivering the final stroke via a lever, divorced the creator from the music, resulting in cheap synthetic sound. Purists have derided new forms of technology with the same accusations of crassness and artificiality long before the TR808 drum machine came along. The times change, but the song – or at least the criticism of it – stays the same.

Cahill, deviser of the Telharmonium, of course saw nothing wrong with creating mechanical replacements for human muscle. To him, it was both an engineering achievement and an unparalleled business opportunity, one worth constructing a 200-tonne organ for. But his gigantic creation was a flop. The real game changer was radio. The ability to transmit sound from one single source to multiple locations was revolutionary. And this time it didn't have to sound like the Telharmonium.

Thomas Edison (1847–1941), Guglielmo Marconi (1874–1974) and Nikola Tesla (1856–1943) were just three of the inventors who chased the idea that sound could be transmitted via radio waves. The precise winner of the race to invent and patent the revolutionary achievement is an emotionally charged subject. American cartoonist Michael Inman of *The Oatmeal* asserts that Edison, a shrewd entrepreneur but not a true geek, stole every good idea he ever had from Tesla.* Inman has gone to heroic efforts to have a museum built to honour the work of Tesla, whom he describes as 'the world's greatest geek'.

---

* http://theoatmeal.com/comics/tesla

Whether or not this is true is another subject, but it is worth noting that radio was not the only audio technology that interested Edison.

## IMMORTALISING SOUND

It is hard today to appreciate – when so many of us can carry 120,000 songs on a wad of metal and plastic no wider than a playing card – that just a century ago all the music everyone ever experienced was transient, temporary and live. Music was something you made or experienced in a specific space and time. It was not something that could be captured, preserved, changed, edited or transported. Today, our capacity to record and transmit sound has reached other galaxies: the Brandenburg Concerto No. 2 by Johann Sebastian Bach (1685–1750) is immortalised on a record made of gold,[22] launched on the NASA Voyager probe in 1976.[23] It has now left our solar system, and could well be enjoyed by alien ears.

Numerous experiments in recording sound took place at the end of the nineteenth century, the most popular of which for a time was the wax cylinder phonograph devised by Edison, the earliest commercial medium for recorded music. Heavy cylinders made of wax bore the etched imprint of a song, read out by a rotating needle that transmitted the vibrations to a broad horn that amplified the information. Earlier prototypes were made with tin foil woven over a hand-cranked core.

Cylinders enjoyed popularity for about thirty years until 1915, when round discs cut with a stylus devised by American inventor Emile Berliner – dubbed 'gramophones' to demarcate them from Edison's cylinders – hit the scene. The original records were only five inches in diameter, and needed to be rotated by hand. Though regarded as a curiosity, discs grew both in size and popularity throughout the first half of the twentieth century. Despite all predictions that the medium would die, vinyl records are with us still.

Another key plug in the evolution of music came from legendary inventor Lee de Forest (1873–1961) and his creation of the audion, a vacuum (airless) glass tube that allowed for acoustic signals taken from one source to be controlled and amplified by another. Rather than having to be laboriously cranked by hand, audions allowed musical volume to be effortlessly turned up.

With this, he set us on the road that we continue down in pursuit of the loudest sounds we can possibly make. An insatiable desire to make our music bigger and bolder seems to define our species: just look at the enormous Taiko drums of Japan and thudding sound systems of German industrial nightclubs. The need for noise is universal.

With a bit of imagination, even the most lo-fi instrument can be impressively amplified. With none of our modern electronic tools at his disposal, American guitarist and one-man musical library Huddie William Ledbetter, or Leadbelly (1888–1949) simply doubled up every string on his guitar to maximise his audio output ('and because it made the ladies dance').

Things might be too loud today, as audiophile critics of the 'loudness war' that has raged in the music industry for seven years contest. With the increasing sound levels that can be attained through electrical engineering and technical manipulation, subtle textures and low-frequency details are sacrificed on the altar of volume. Adolescent tinnitus and illegal warehouse raves probably weren't what de Forest had in mind when he created his thermionic valve, but he played a crucial part in their creation nonetheless.

# MAGNETIC MUSIC

One medium which perhaps more than any other changed forever the way we make music may surprise the younger readers of this book: magnetic tape. Though the shiny black ribbons are alien to children who have grown up in the world of invisible MP3s and shiny DVDs, this material is more versatile than you might think: electronic experimentalist David Vorhaus, for example, made a bass guitar using magnetic tape as the strings, dubbed the kaleidophone.

Eccentric experiments aside, we might never have developed the capacity to edit and manipulate sound waves were it not for magnetic tape. All music would have been recorded faithfully as a documentation of the live performance of an artist.

Though less robust than vinyl, magnetic tape allowed musicians to experiment and play with sound in ways they never could before. On the weirder end of the spectrum, members of the French avant-garde *musique concrete* movement pioneered the 'cut and paste' approach that is ubiquitous among today's samplers and mash-up artists. Back then, without the ctrl-x function on their laptops, experimentalists had to use scissors and Sellotape to cut and paste magnetic tape back together. In their quest to impose order on chaos, they manipulated and distorted sounds in every way they could think of, such as playing the tape backwards, employing sine wave generators and analogue sequences (tools co-opted from the laboratory for the musical studio), and mixing classical music with electronic sounds.

They were also renowned for their weird use of 'found sounds', such as the tapping of pans or random industrial noises, an idea that harks back to 1913 and Italian Luigi Russolo (1883–1947), widely regarded as the first 'noise artist'. He asserted in his 1913

51. The audion, composed of bits of wire strung through a vacuum chamber (sucked free of all gas), allowed for electrical signals from one source to be amplified by another. Result: cranked volumes, soundsystems, and rock and roll as we know it.

manifesto *The Art of Noises* that the industrial revolution had given our species an unprecedented sonic landscape and therefore musicians had been gifted a new acoustic palette. The techniques of the *musique concrete* movement would later be embraced by a number of artists who disseminated their strange techniques to the mainstream – most famously, the BBC Radiophonic Workshop, best known for creating the *Dr Who* theme in 1963.

Most importantly, magnetic tape allowed for one revolutionary change that, upon reflection, seems so simple yet at the time sounded unimaginably absurd: the capacity to record one sound on top of another. The first to do so was the American electronic engineer and whizz-kid Les Paul (1915–2009), creator of one of the first electric guitars. His all-night experiments with gadgets and gizmos in his garage were integral to the history of rock and roll. The world owes Paul a great debt, for it was he who came up with the idea of multitrack recording: the capacity to record sound waves separately and then combine them into one stream of music.

## SOUND ON SOUND

This might seem like a fiddly technicality, but multitrack recording meant that session musicians did not have to be in the same place at the same time to produce a mastered track. And it wouldn't have happened if rogue geek Paul hadn't explicitly written to electronics manufacturer Ampex asking if they could add a second head to their magnetic tape recorders (which would allow musicians to listen to the track as they added to it). The idea at the time seemed bizarre. But, without it, we would have been chained to the compressed vinyl products that immortalised sounds as they were made, and the entire universe of studio recording, electronic music, and now the studio-in-a-laptop would not exist.

From the 1940s onwards the studio increasingly became the focus for musical creation, as engineers developed the capacity to meticulously rerecord and polish individual streams from each member of an ensemble. Musicians often were forced to challenge the overly cautious nature of recording engineers in order for their creations to evolve. British studios were especially renowned for their conservatism in comparison to their American counterparts. Engineers at Abbey Road took their profession so seriously, they wore lab coats to minimise the chances of leaving dust or bacteria on the equipment (as well as to symbolise the importance and seriousness of their profession).

52. We don't call them sound 'engineers' for nothing: Abbey Road's technicians would don lab coats not only to minimise the chances of depositing hair and dust on the equipment, but also to emphasise the scientific gravitas of their work.

George Martin (the fifth Beatle) had to twist many arms to convince them to try outlandish new techniques that today are routine, such as putting a microphone into the drum kit.

## SYNTHESISED SOUNDS

Seventy years after the gigantic Telharmonium Mark II drifted to New York in a fleet of cargo ships, sound engineers returned to the idea, creating pure unadulterated tones through electronic means alone.

The 1970s and 1980s saw the introduction of synthesisers, a range of devices forged in the laboratories of electronics manufacturers and co-opted by a new generation of musicians who could, for the first time, replicate a recording studio at home. These allowed music to be made by a wider range of people because they were cheap, portable, and did not require formal education. In other words, sophisticated electronic instrumentation was disseminated to the masses.

53. The TR 808, crafted in the demure labs of Japan for the
karaoke market, co-opted by America's hip hop artists for
the curation of the sickest basslines on the planet.

The TR303 and TR808 synthesisers – both produced by
Roland – are two of the best-known examples. Musicians who
could not afford a bass guitar or to hire a bassist for their recording sessions could suddenly replicate the sounds they wanted
with a machine that would fit on a chair. If you live in a city,
you hear the sounds of the TR808 every single day. It forms the
bedrock of bass lines for almost all contemporary hip hop and is
universally praised through devotional lyrics as equally fervent in
their worship as evangelical gospel hymns. Teens whose parents
could not afford expensive music lessons could suddenly bring
their sonic visions alive inside tiny bedrooms. A device devised
in the demure laboratories of Japan, intended for the Karaoke
market, has become the basis for gangsta rap. Probably not what
its Japanese inventors intended.

The TR 303 drum machine was equally important for the
development of house music, and gave birth to the culture of

Ibiza. Now another form of technology is supplanting them all: the laptop computer, with programs such as Garage Band – an entire studio in a machine the size of a magazine.

Enter the Internet, and now musicians can create new forms of music collaboratively with people they've never met, giving their scores unprecedented forms. The evolution continues.

Technologists, inventors, engineers, artists and scientists of varied stripes have continually given us new ways to make music. Yet one gift, more than any other born in the laboratories of the record industry, has changed our understanding of what music truly is: the development of brain scanners.

Former record industry titan the EMI Group once held Parlophone, Capitol Records and EMI Records in their hoard, and prior to their downfall in 2012 were one of the 'big four' labels. But before they brought us the Beatles and Michael Jackson, Electrical and Musical Industries Ltd (formed in 1931) was an engine of scientific innovation, not popular music. EMI's laboratories produced an impressive array of technologies, from the radar detectors and microwave devices used by the Allies in the Second World War to the BBC's first television transmitter and early colour cathode ray tube TV sets.

EMI funded the research that led to the creation of CAT scanners (also known as X-ray CT scanners, for computer tomography).[24] These were known for a time simply as 'EMI scanners', and revolutionised our capacity to photograph our internal anatomy. This won EMI the Queen's Award for Technological Innovation in 1970 and engineer Sir Godfrey Newbold Hounsfield (1919–2004) the Nobel Prize in physiology or medicine in 1979. Hounsfield realised computers could be used to mathematically analyse X-rays and produce maps of the human body. The day he threw the switch he launched the modern era of neuroimaging.

Hounsfield wanted to follow this line of thinking in the early 1950s. But it was not until the 1960s that EMI was willing to fund his expensive research. It has been argued that the record company was only able to do so thanks to the Beatles and the unprecedented profits they reaped for the record company.

Do we have the fab four – and rock and roll itself – to thank for the creation of the scanners that have illuminated how the human brain works? Possibly. At the very least, rock and roll sped up scientific progress and brought us to where we are today in our understanding of the mind.

## TURN UP THE BASS

Of all the things that humans do, music must be the strangest. The power of sound is potent and pervasive. So much so, that even for those who experience the world in silence, music can still intoxicate and enchant.

I grew up in the music industry, worked at a nightclub for five years, and with Guerilla Science have operated venues at dozens of music festivals, including a rat maze next to the Hell Stage in the middle of Glastonbury's 'naughty corner'. I've seen grown men moved to tears by the beauty of an opera, heavy metal crowds maniacally crush towards the stage to the point of self-obliteration, and felt the uplifting joy at countless gigs as the audience is carried to a new plane of camaraderie. Music has a strange effect on our strange species.

But on one unforgettable occasion, the power of music became more apparent than it ever has to me – through a sight, not through a sound.

It was dark, and it was loud. Incomparably loud. With nobody to chat to (and not speaking the language of the crowd), I decided to make friends the only way I could. I tapped a friendly looking stranger on the shoulder. He turned around, regarding me quizzically as I held up a pink balloon.*

I inflated, tied it, and handed it to him. Wrapping his hands around the pink bubble, his eyes widened slightly as he read the inscription, and a smile spread over his face. He nodded at me approvingly, and wandered off with his acoustic bounty.

Emblazoned with FREE BASS in bold white lettering, the balloon was one of dozens handed out that evening. Everywhere hands gripped them tightly. Elsewhere other hands punched the air, in time to the thumping music wafting from the towers of speakers. Some hands wrapped around those of others in alluring, friendly invitations.

But most hands were doing something you rarely see: dancing through the air, darting this way and that in silent, animated conversations. Everywhere friends and strangers chatted in silent, passionate, expressive discourse. Sometimes from across the hall, hundreds of feet away from each other, unperturbed by the overpowering bass.

'I needed a new challenge – I was tired of being "cool" and putting on big, standard-issue club nights. I wanted to do something new. So I asked myself: in the world of music, what is impossible?' says Ronald Ligtenberg, founder of Sencity, club nights designed to bring deaf and hearing people together through a mutual love of music.

Born in the Netherlands, Sencity has gone to Spain, Finland, Brazil, Mexico and South Africa before coming to London's

---

* I cannot count the number of times I have done this at music festivals and people nearby thought that I was selling nitrous oxide balloons and instantly lined up proffering cash.

o2 arena in 2011 – a surreally conventional setting for such an unusual event. Inside the airy, sterilised, shiny dome, more like a shopping mall than concert hall, 2,000 people gathered for an event unlike any ever heard in the UK.

If you have ever held a balloon inside a nightclub (or muddy field-cum-party), you may have discovered that it vibrates with music. The sac of air acts as a natural amplifier for sound waves, and low frequencies in particular. Hence the pun: 'free base' is the nickname for a cooked alkaloid, such as 'free-based speed', or roasted amphetamine.

Guerilla Science has doled these balloons out for years, and we have found that people tend to be split in their reaction to them. People either love them or completely fail to understand what they do. Sometimes they stare at us in quizzical bemusement. But, most of the time, people instantly understand, no explanation required.*

Charmed and delighted, they request several more, clutch them to their chests, and find ways to enjoy them that even we had not thought of. Some have pressed together a string of balloons four or five in series, enjoying the bouncing chain of sound.†

Others have found ways to trap them inside the grates of sound systems (technically termed 'bass bins'), watching the pink spheres clatter in time to the beat. Couples have held one between their heads, enjoying vibrations in mutual appreciation.

It isn't hard to imagine why: deprived of the capacity to appreciate music through their ears, it is the physical nature of music that matters most.

---

* http://www.flickr.com/photos/guerillascience/4927196198/in/
set-72157624681717313
† http://www.flickr.com/photos/guerillascience/4926600305/in/
set-72157624681717313

54. Fancy some free bass? I cannot count the number of times people thought we were actually selling free-based amphetamines. The pun never ceases to amuse.

Music, we sometimes forget, is a tactile sensation as well as an auditory one: our experience derives from feeling sound as well as hearing it. The sensation of hearing itself, the pattering of sound waves on the bouncy membrane of the ear drum, is 'touch at a distance' as British-born American perceptual and cognitive psychologist Professor Diana Deutsch puts it.

But once it reaches the brain – whether through the ears or the body – music can hijack our internal hormonal messaging systems and the electrical highways of the body to produce bizarre experiences that are felt from head to toe. Music can send shivers down the spine, raise goosebumps on skin, and even cause a vibrating thrum in the heart, just like the rush of being in love.

Nights that cater for the non-hearing take the tactile nature of music to its fullest expression, with massive sound systems blasting the heaviest basslines possible. Sencity even features a vibrating dance floor that thumps in time to the music. It has travelled the world over.

At these nights, 35 per cent of the audience is 'profoundly' deaf, and another third hearing-impaired in some way. Ligtenberg couldn't put it better: 'A third of the people here cannot "hear" at all. But they still love music.'

## PISCINE ORIGINS

That music can still be felt, appreciated and devoured by people who do not perceive it through their ears strikes at the core of what music truly is: not an artefact of sound, but a construct of our brains.

In fact, 'sound' itself is a construct of our brains. This harks back to the old philosophical puzzle: if a tree falls in the forest and nobody is there to hear it, does it make a sound? The answer is straightforward: No. Vibrations between molecules in the air exist. Mechanical waves, like the motion of the ocean, propagate through the air from the thud of the tree. Molecules bump into each other, onwards and outwards. This is why the explosion of an atomic bomb can flatten buildings and trees far beyond the point of impact: sound waves, if sufficiently strong, can move with the force of a tsunami.

It is not until those waves reach your eardrum and that mechanical motion converts to sensations within your brain that 'sound' actually exists. Until then, it's nothing more than the silent jostling of airborne molecules.

This explains why people who cannot hear still throw raves: music has tactile qualities as well as auditory ones. And why their

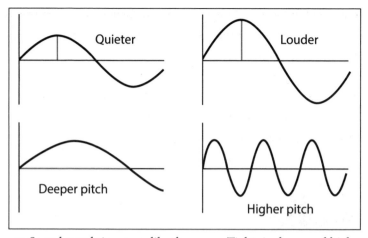

55. Sound travels in waves, like the ocean. Tight ripples sound high, slow undulations low. Tall crests are loud, and gentle waves soft.

parties are so deliciously loud: the more pressure there is behind the sound waves, the more they can feel the music.

For all of us, there are sounds we cannot hear: the human hearing range varies from 20 to 20,000 hertz, hertz being a sonic unit of measurement of pitch, or frequency. This is measured by the distance between the peaks and valleys of sound waves: the tighter the ripples, the higher the note. The higher the wave, the louder the volume.

Sounds higher in pitch than 20,000 hertz are known as 'ultra-sounds', such as those used in medicine to create images of our innards – like the echo-locating skills of bats. Just as ships use sonar to detect enemy submarines, we can use sound to make maps of our anatomy. A wide variety of animals navigate their world and communicate with sounds that are too high in frequency for us to hear, such as bats, rats, dolphins and frogs. At the other end of the scale, some animals communicate with sounds too low for us to hear, such as whales and elephants.

Why should low-frequency bass tones feature so strongly in

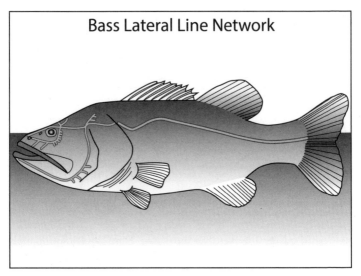

**Bass Lateral Line Network**

56. Fish have another anatomical structure that they use to sense sound: the lateral line network, here illustrated on a sea bass.

dark, loud clubs? The answer appears to lie in ancestors even older than those who made the ochre markings on cave walls: fish.

The auditory receptors of the animal kingdom come in a vast array of shapes and forms. Rabbits, bats and elephants all bear spectacular extracranial attire. Cricket ears, fairly simple drum-like membranes, are located on their legs. Fish hear not just through their 'ears' (termed otoliths), but also through two threads of sound-sensitive tissue, dubbed 'lateral' lines. These allow fish and sharks to three-dimensionally sense motion in the water around them: spectacularly sensitive biological strips of 3D headphones.

Within the otoliths of fish lies their main acoustic anatomical utensil: the sacculus, a fluid-filled bag of sound-sensitive cells. This structure was retained within the skulls of animals as they migrated on to land, and remains in place in amphibians, reptiles,

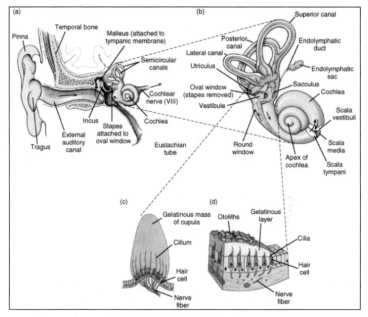

57. The sacculus is an ancient, Piscean structure that resides still deep within our heads, snuggled next to the modern contraptions we use for hearing.

birds and mammals, though it is thought to play a role in hearing only in fish. In terrestrial animals, such as ourselves, the sacculus is thought to now only be used for balance and spatial navigation as part of the 'vestibular system'. It is this portion of your anatomy that produces dizziness if you have an inner ear infection.

But in 2000, auditory scientists discovered that, in humans, this ancestral piscean organ responds to sound. And not just any kind of sound.[25] Tests with people revealed that our sacculus is most responsive to tones lying between 300 and 350 hertz – within the range of music (middle C is 261 hertz). The sacculus only responds to music above 90 decibels: loud. The kind commonplace at rock clubs and raves.

Moreover, there are direct connections between the vestibular system and a region of the brain called the hypothalamus, a small but powerful part of the brain that sits roughly in the centre of the skull, right above the troublesome pituitary gland, the mischief-maker of puberty.

The hypothalamus is one of the main links between the wires and neurotransmitters of the nervous system and the oozing organs and molecular hormone messengers of the endocrine system, and it plays a key role in regulating basic impulses such as hunger, thirst, fatigue – and sex. Our ears are the most basic link between sex and rock and roll. Moreover, the direct physical links between our means of maintaining balance, sense of hearing, and the hormones of feeling could help explain the evolution of dancing.

Work at the Virginia Tech Carilion Research Institute reveals that neurons might release neurotransmitters in response to sound, and to low frequencies in particular. Professor William 'Jamie' Tyler is working on ways to treat brain injury with low-frequency bass notes,[26] reviving an old idea that the brain is a mechanical machine as well as an organic organ. How did he stumble upon this idea? By listening to loud music late at night as an undergraduate and noticing 'spikes' in the activity of dissected neurons on his lab bench every time his subwoofer boomed. Musical recreation led to scientific innovation.

The irresistible hedonistic drive to turn up the bass inspired an insight that could not only change what we know about how the brain works, but aids our capacity to heal those who have suffered damage to the centre of their cognitive being. There may yet be more value – intellectually, scientifically and medically – to be found by cranking up the volume. There is redemptive power in making cacaphonous noise.

# RELIC BONES

Moving out from the ancient depths of our inner ears, let us probe the rest of the fleshy microphones that form our auditory windows on the world. The weird wriggly bit on the outside of our heads is just one of three components of the ear. Called the 'pinna', the outer ear evolved to channel sound waves into the head. The curvature and shape of each is so distinctive that it could be used as a biometric form of identification, just like a fingerprint or scan of the iris (the colourful region encircling the black pupil in the centre of your eye).

Sound waves travel down the auditory canal to the bones of the middle ear. The eardrum, 'tympanum', is a thin membrane stretched across the cradle formed by the first of three tiny bones, called the 'malleus'. This translates as 'hammer' (a rather rock and roll nickname). The malleus transmits sound vibrations to the next bone in the series, the incus, or 'anvil'.* This in turn rattles the last tiny bone in the middle ear, the stapes, or 'stirrup', thusly named because it resembles a saddle.

These three bones – together termed the ossicles – are the smallest bones in the human body, and the stapes is the very smallest of all. It snuggles up to the cochlea, the sensory spiral of our inner ear.

Quirkily, the middle ear bones did not used to live in the ear: they used to reside in the jaws of our ancient pre-mammalian ancestors. In reptiles and amphibians one single bone, the

---

* On that note: If you ever feel creatively sapped or emotionly destitute, devote 90 minutes to the documentary *Anvil* (2008), which tells the undeniably endearing story of Canadian heavy metal band Anvil, who haven't stopped gigging or believing since their brief brush with fame in the 1980s. I promise: time well spent.

'columella', connects the eardrum to the inner ear.

Two bones from the joint of the jaw – the 'articular' and the 'quadrate' – migrated into the middle ear, forming this new chain.

This is one of the best-documented examples of what evolutionary biologists call an 'exaptation': the repurposing of an existing structure through the course of evolution.

The bones of the middle ear are one of the features that unites all mammals. German anatomist Karl Bogislaus Reichert (1811–1883) first established the similarities between mammalian ear bones and reptilian jaw bones based on embryological evidence in 1837. Charles Darwin's *Origin of Species* was published two decades later, in 1859, just to put that in perspective. Establishment of this connection is considered to be a milestone in the history of science and our understanding of evolution. Species change into others over time. This idea changed the world.

While the bones of the middle ear migrated from the jaws of reptiles into the ears of mammals, in whales, the opposite happened. Their hearing apparatuses changed dramatically when the animals moved from dog-like land-living creatures into water-dwelling cetaceans. Sound waves travel farther in water with different acoustic properties, so whale ears had to adjust. Now the bone that houses the inner ear – which is enormous – is located in the jaw, where sound waves can more easily reach the spiral of the cochlea by travelling through the fatty tissue of the throat.

A recent study[27] hints that the size and shape of the human middle ear hammer changed early in our evolution and could even be regarded as a defining characteristic of humanity. This is the firm opinion of palaeoanthropologist Rolf Quam of Binghamton University in New York State, who based this on an examination of the ossicles from a 1.8-million-year-old specimen of the ancient hominin *Paranthropus robustus*.

Professor Quam then compared these wiggly bones to a

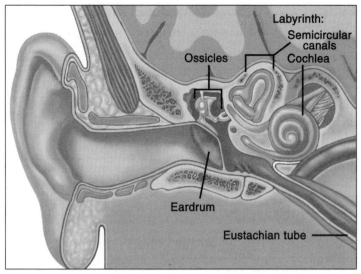

58. The cochlea, the spiral of the inner ear, undeniably resembles the shell of a snail. This is the final destination for sound waves before they are processed into signals for the brain.

set from its ancestor, *Australopithecus africanus*, which lived between 3.3 million and 2.1 million years ago. The modern human hammer, the 'malleus', is small compared to that of chimps and other apes. *P. robustus* and *A. africanus* however both had small malleus bones, like us, which would translate into a different hearing range for our ancestors.

Other studies[28] have shown that the relative size and shape of middle ear bones of other apes correlate with what range of sounds they can hear. If you're wondering how they do this, the answer is charming: biologists can train animals to indicate the softest sound they can hear for any given pitch. For example, an ape or a gorilla raises its arms to indicate if it hears a sound. Biologists can produce an 'audiogram' for a given species in this way, showing the range of sounds a species can hear.

Audiograms for a wide range of animals have been produced

in this fashion, from horses[29] to turtles.[30] Working with the latter must have required a remarkable level of patience, even for a laboratory scientist.

Professor Quam thinks that the shape of the malleus has a 'deep and ancient origin' and could be, like bipedalism, a defining characteristic of our species.[31] His assertion for the evolutionary significance of the malleus is not without controversy. Others in the field consider his extrapolations overblown. But the intricate structure and evolutionary significance of the middle ear bones remain intriguing.

What is without controversy is the well-established history of our middle ear bones: they are the smallest in the human body, and because of their intricate structure and biomechanical delicacy, grow very little as we age. The lens of our eye does the same, which is why babies have such massive peepers compared to adults. Biologists thus have a special name for the jiggly bones of the middle ear: 'relic bones'.

There is another mind-boggling feature of the microphones inside your head. The ear-drum vibrates in and out as molecules hit it, like balls tossed on to a taught piece of rubber. Each molecule brings with it a different set of information about where it has come from (the location of the sound), the speed at which it hits the drum (the volume), and its vibrational frequency (the pitch). Here's the reptilian jaw-dropping, mammalian hair-raising thing: if the ear drum is bombarded with millions of different molecules at any given time, how on earth does it manage to transmit the information from each individual particle to the ossicles and on to the cochlea? Imagine a thousand balls pummeling the taut rubber at once.[32] How can the eardrum tease apart the information from all of them? The ear manages – and the mind boggles. Our tympanum kindly accomplishes this, every moment, every day, without us even having to think about it.

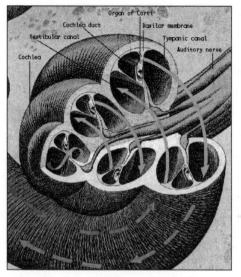

59. Along the length of the spiral of our inner ear runs the Organ of Corti, which bristles with hairy spikes that shiver in response to sound.

## SONIC SPIRAL STAIRCASE

Moving deeper into the skull, the plot thickens when we reach the spiral of the inner ear, the cochlea. The stapes – the last in the chain of three ear bones – connects to an opening at the top end of the spiral, called the oval window, where it presses inwards and outwards like the piston of an engine. In so doing, it compresses the fluid filling the spiral, gently squishing it in and out. Along the length of the spiral runs a ribbon, the basilar membrane. And along this membrane runs a thread of sensory cells, called the 'organ of Corti', so named after the nineteenth-century Italian contemporary of Reichert, Marquis Alfonso Giacomo Gaspare Corti (1822–1876).*

---

* I find it amusing that it is called an 'organ'.

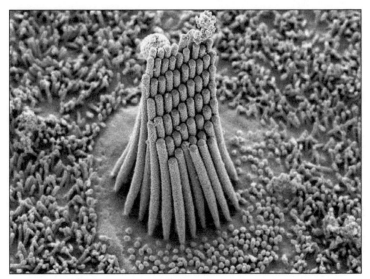

60. The hair cells of the inner ear possess what must be one of the loveliest biological names: stereocilia. Each bundle quivers in sympathy with a different frequency of sound.

The organ of Corti – like the middle ear bone ossicles, unique to mammals[*] – bristles along its length with the spectacularly named 'stereocilia', also known as 'hair cells'. These gather together to form tight bundles at regular intervals along the basilar membrane, like keys along a piano.

Though I have tried in this book to describe the spectacular shapes of biological structures with words that do them justice, sometimes words are not enough. Have a look.

These bundles of cells are each triggered by a different frequency of sound based on the thickness of the basilar membrane that they spring from. The membrane varies in thickness from the top to the bottom of the spiral of the inner ear. It is thickest at the

---

[*] Except for monotremes – mammals that lay eggs, such as the platypus and the echidna (the latter of which you might recall has a startling bifurcated penis).

top: this portion only moves in response to high-pitched notes. Sound waves with the highest frequencies produce movement at the top of the membrane, jiggling and stimulating hair cells, which then transmit their movements on to sensory nerves – the electrical wires of the body. These then conduct their message into the brain, where it registers as a high-pitched sound. Low-frequency bass notes travel further down the membrane to the bottom of the cochlea where the membrane is thinnest, setting off the stereocilia of the inner ear's ocean floor. Every minute, every day, the spiral of your inner ear continually rolls like incessant waves surging onto the seashore.

When I described the overall construction of the eardrum, middle ear bones, and the spiral of the inner ear to my mother, she remarked, 'So it's shake (tympanum), rattle (ossicles), and roll (cochlea)?' I wish I had made the observation myself – but I would never plagiarise, especially from the woman who gave me my ossicles. I quote her quip with gratitude.

From the spiral of the cochlea, neurons threading from each bundle of hair cells gather together into the auditory nerve, which faithfully transmits their signals to a horizontal strip of the brain, the auditory cortex. This sits conveniently close to the ear in the outermost layer of the brain, and is patterned in what is called a 'tonotopic map': the highest-frequency notes are received at the front, closer to your face, and the lowest-frequency notes closer to the back of the head.

Unlike the somatosensory cortex, which we met at the start of this book when scientists mapped the sexy bits of the brain, the auditory cortex – the map of sound in the brain – is a direct translation: stripes of the brain respond to pitches in sound in a straightforward descending fashion, like a staircase.

This tiny region of tissue, no bigger than your little finger, is where the real musical performance begins: the brain.

# ORGANISED SOUND

Humans do many strange things – and few could be more baffling than music. It is both beautiful and bizarre that we can classify symphonies according to major and minor keys, describe one string of notes as 'happy' and another as 'sad', and discuss the expressive quality of music as though frequencies – the mere jostling of particles in the air – have emotional meaning. How can one set of notes sound angry but another joyful?

It is even stranger that music should exist in the very first place; we can hardly even say what it is. To define or even describe the strange phenomenon that is music through words is uniquely difficult. Mostly we only realise how tricky it is to really define music when we are forced to try. It is astoundingly challenging to identify this ubiquitous entity we have encountered throughout our lives.

A fair and sensible definition is 'organised sound'. But countless anthropogenic, or man-made, sounds are organised, and cannot be categorised as musical. Would an alarm clock, police siren, or ringing phone qualify as symphonic?

Philosopher (and clarinettist) David Rothenberg, author of *Why Birds Sing* (2005) describes music as 'organised sound, for its own sake'. This advances us slightly in our quest for a fitting definition of music – particularly, the 'own sake' qualifier. Sound created for the pure joy of producing sound, with no other apparent benefit. That sounds like music. And yet still something lacks. Shouting, chanting and screaming are all done for the pure joy of producing sound, but they are not always musical.

A more alluring definition is 'beautiful math'. Music is indeed mathematical, from the balanced ratios of harmonics to the regular beat of a rhythm. And it most certainly can be beautiful. But many forms of mathematics are beautiful. Those of us who

61. What is music? There are many definitions. One of the most convincing: 'beautiful math'. But countless biological forms are mathematical without being musical.

have never enjoyed playing with numbers can nonetheless appreciate the harmonious, symmetric, elegant beauty of geometric shapes, trigonometric patterns, and fractal explosions. Biological forms bloom in millions of mathematical shapes, from the spirals of snail shells to the branching veins of leaves. The biological forms are mathematical, but they are not musical.

Moreover, music is not always harmonic, symmetrical or mathematical. There are archaic, anarchic, emotive, rebellious elements that prove that what to one generation sounds 'bad' to the next can sound 'good'. New forms productively break old. Music is always evolving, changing, shifting and moving. An infinite variety of forms are available to it. Musical innovators, from jazz saxophonists to the avant-garde composers of the 1960s to punk rock noise makers and today's eccentric 'found sound' experimentalists, from Penderecki to Terry Riley and Matmos, are proof of the redemption in rebellion. Music need not be constrained by classical mathematical harmonies to strike a chord with audiences. Music cannot and should not allow itself to be constrained to any neat box. The history of music is a history of revolution and reformation. Frequently aided by our familiar friends invention, imagination, technology – and drugs.

And musical rebellions – even the strangest combinations of chirping and clapping,* which to many do not qualify as 'music' – are justified as 'musical' by recent scientific discoveries. We need look no further than Brian Eno for the clearest explanation of their intentions:

> So if this was 'experimental music', what was the experiment? Perhaps it was the continual re-asking of the question 'what also could music be?' … It moved the site of music from 'out there' to 'in here'. If there is a lasting message from experimental music it is this: music is something your mind does.

Repeat: 'Music is something your mind does.'

If weirdo avant-garde experiments gave us nothing but the illustration that music is a construct of the mind and not an artefact of the instrument, that would be sufficient.

## AN EXQUISITE ILLUSION

So what, then, is music? The most salient philosophical definition that could describe it is challenging and only makes sense in the light of neuroscience: 'an exquisite illusion'.

---

* See: Meredith Monk, one of the movement's most eccentric artists.

Historically, almost all philosophical and scientific attempts to understand music focused on its geometric and mathematical qualities. For example, though the number of notes in a musical scale varies between cultures, the perfect octave – a doubling of frequency (such as C4, or middle C, and C5, tenor C) – appears to be universal. The tritone on the other hand – a combination of three notes deemed so unpleasant that it was known as *diabolus in musica*, 'the devil in music' – is so jarring that even babies instinctively recognise it.[33]

It was not until the late twentieth century, when neuroscience and scrutiny of the machinations of the mind flourished, that we began to see music for what it truly is: a product of the mind.*

It is a figment of our imagination. Music exists in our heads. And it exists only in our heads. This is also true of the nature of sound itself. But sound waves have a direct relationship to the vibrational frequencies of molecules propagating through air. What we hear truly relates to a measurable concrete property of the physical world around us.

Music is another beast entirely. It does not exist in the objective world around us. It is a subjective reality that our brains create for us, confined to the insides of our skulls. As neuroscientists increasingly appreciate, our brains are pattern-making machines, and music is the ultimate pattern.

'The brain must draw on what it can in order to understand the sounds that are presented to it,'[34] explains Dr Diana Deutsch of the University of California in San Diego. British-born and transplanted to America, she is known as the 'grand dame of acoustic illusions'. She has compiled an unparalleled range of

---

* As we noted in the previous chapter, Brian Eno stated 'music is something your mind does', proving yet again that artistic insights can precede scientific revelations, and that Brian Eno is a dude.

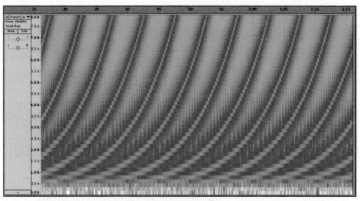

62. A spectrogram of the Shepard-Risset Glissando, which has been dubbed the sonic equivalent of a barber shop pole, forever ascending and descending like an Escher staircase.

trippy auditory tricks: strange sonic teases which acoustically illustrate that the mind makes what we perceive.

Take for example the Shepard Scale, also called the 'never-ending scale',[35] devised by cognitive scientist Roger Shepard (1929–1987). Listen to a recording online, and you will find that it sounds as though the notes are forever increasing, when they are actually circling round in sonic rings.* Software overlaps the first and the last note in the scale to give the impression that the sequence has come back to the beginning. It has been dubbed the auditory equivalent of an Escher staircase. The Beatles embedded a Shepard scale into the final throes of 'I Am the Walrus'.

This effect is taken to its fullest expression in a *glissando* – Italian for 'sliding' – developed by French composer Jean-Claude Risset. The 'Shepard–Risset glissando' – likened to a sonic barber shop pole – is used to spectacular effect by eccentric Icelandic

---

* You can hear a version – and several other auditory illusions – in the Guerilla Science Sonic Tour of the Brain, an 18-minute audio tour of what the brain sounds like: http://guerillascience.co.uk/archives/3750

pop princess Bjork on her album *Biophilia*, using human voices instead of computers.

Auditory illusions which exploit our brain's habit of filling in the blanks and hearing things that are not there are not unique to the modern age of auditory physics and electronic music. Classical composers in Sardinia found that by combining four male singers in just the right fashion, a female soprano suddenly seems to emerge. Known as *la quintina*, meaning 'the fifth one', it was said to be the voice of the Virgin Mary blessing the singers for their vestal virtuoso. Musicians around the world have achieved the same effect, but the Mediterranean version is the best documented, studied by French scientist Bernard Lortat-Jabob.

These and other illusions illustrate that we don't 'hear' the world as it is. We perceive the pattern our brain can best impose on the information it receives. This is a difficult concept to grasp. Even the most experienced neuroscientists who study music still struggle daily with it. But grappling with this uncomfortable truth is the only way to understand and appreciate the phenomenon that is music.

Yet the fact that music is an illusion does not render it insignificant. Quite the opposite: it makes it even more remarkable.

## NO MUSIC

The best (and perhaps the only) way to illustrate the illusory nature of music is by introducing the very strange yet surprisingly common condition known as *amusia*, meaning 'no music'.

'Suffer' may not be the appropriate word, but people who are afflicted with amusia – so-called amusics – are unable to hear music. The anatomy of their ears is usually normal, their hearing tends to be perfectly functional, and by and large they possess no other abnormal neurological traits. They tend to have unremarkable mathematical and linguistic capacities. There usually seems to be nothing whatsoever strange about them. Their intelligence, memory, and other cognitive capacities are typically average. Most are not even aware that there may be anything unusual about them. In fact, they are so unremarkable in every other way that we only recently discovered how common they are: possibly up to 4 per cent of the human population.

They do not hear what most of us do. When most of us hear a swirling crescendo of violins, structured harmonies that sound pleasing and fitting, amusics hear disjointed noise. Amusia is more than simply being 'tone deaf' (a bashful way of admitting to an inability to sing). The difference between notes, which the rest of us are primed to notice, is imperceptible. While the deaf can throw club nights, the hearing can detest the sound of a cello.

We will, of course, never know for sure, but it has been postulated that a few notable historical figures were amusic – most appropriately, the world's most famous guerrilla, Che Guevara, who apparently couldn't distinguish a samba from a salsa. It has also been suggested that Sigmund Freud was amusic, as well as America's 18th president Ulysses S. Grant (1822–1885).

The first formal record of 'note deafness' appears to date to 1878, but it was not until the advent of MRI scanners and electroencephalography (EEG) sensors that scientists could begin to identify which parts of the brain were behaving atypically, and why.

Psychologist, cognitive scientist and cellist Professor Isabelle Peretz studies amusia at McGill University in Montreal, Canada.[36] Dr Peretz was the first scientist to publish a formal description[37] of 'congenital' amusia – meaning present from birth – in a woman named Monica. Monica's case was so severe that she could not tell if one note played after another was of a higher or lower pitch, even if the two notes were spaced far apart on the scale.

Amusia affects roughly 4 per cent of the population, Dr Peretz says. By comparison, 7 per cent of men are colour-blind. 'Perfect pitch' is far better known than amusia, but only one person in 10,000 has truly perfect pitch (or as it is technically known, 'absolute pitch', the ability to name any note when played in isolation).

Amusia comes in more than one form – in fact, more than a dozen forms have been identified. For some, music is irritating and even painful – harmonious concertos sound like the crashing of pots and pans on the floor (hence the nickname for such amusics, 'clatterers'). About half of amusics have trouble with rhythm perception, says Peretz.

But others who are still able to perceive rhythm, but cannot perceive melody or the textural contrast between instruments, often enjoy listening to music nonetheless because they appreciate tempo and percussion. The difference in the way a flute and a trumpet sound when producing the same note is referred to as the 'timbre' of the sound. Hence the term for these amusics – 'dystimbrics'[38].

But most amusics, says Peretz, simply find music confusing. For many, in fact, the attraction the rest of us have for these strange patterns of noise is so bewildering, and their exclusion from it so isolating, they are often too embarrassed to admit their confusion. But they will rarely feel alone – and will usually know others who feel the way they do: the condition tends to run in families. Researchers are identifying the genes responsible for certain kinds of amusia, and the specific brain regions affected.[39]

However, amusia may also develop later in life through stroke or brain injury[40] – you too could one day cease to hear music.

# 'EVOLUTIONARY PARASITE'

Peretz is one of the field's pioneers: she has studied how music affects the brain for more than 20 years. In the 1980s, she found it difficult to gain acceptance for her assertion that music is an important biological adaptation. Though it may seem obvious to many that music is an integral component of the human composition, few neuroscientists shared that view 30 years ago. 'I think this was because people perceived music just as a cultural product and a form of entertainment – as 'fun'," she says. 'They didn't take it seriously.'

Most infamously, neuroscientist, linguist, and influential author, Professor Steven Pinker wrote in *How The Mind Works*: 'What benefit could there be to diverting time and energy to making plinking noises? ... As far as biological cause and effect are concerned, music is useless ... it could vanish from our species and the rest of our lifestyle would be virtually unchanged.'[41] This view – in part owing to Pinker's influence on the popular understanding of science – held sway for many years.

Neuroscientists only began to examine the effects of music on the brain with considerable energy and scientific rigour in the 1980s. Prior to this point (and even now) the focus for most neuroscientists remained on language as the hallmark of the human brain: the crucial characteristic that defines our species. Music was considered neither adaptive nor important. Pinker dismissed music as 'auditory cheesecake',[42] a form of sonic junk food that takes advantage of the neural structures in our brains which evolved to process patterns in speech, just as fatty and sugary treats exploit the hardwiring in our tongues that drive us to crave energy-rich foods.

Pinker may now regret this statement, as it has been quoted

derisively and frequently ever since – but he was not alone. More sneering, French cognitive scientist Dan Sperber,[43] Emeritus Director of Research at the Centre National de la Recherche Scientifique (CNRS), dismissed music as 'an evolutionary parasite'.*

The next generation of neuroscientists took issue with the view that music could be anything more than an accidental by-product of the neural networks that produce language. It is no coincidence that many were trained musicians. Peretz is a cellist. Professor Aniruddh Patel in San Diego plays the trumpet. Best known is Daniel Levitin, author of *This Is Your Brain on Music*,[44] who retrained as a scientist later in life following a career as a producer and guitarist.

'Music was not exactly forgotten by neuroscientists, but compared to things like the neuroscience of vision it was neglected. A big reason for this was that the technology to manipulate sound in a controlled fashion did not exist until we had the digital recording technologies that we have now,' he explains.[45] In other words, neuroscientists know what they do now thanks to the work of audio engineers and recording artists. 'Now that we've been able to take a closer look: Music is so much more complicated than I could ever have imagined.'

History is instructive, and no case could be more noteworthy than that of Vissarion Shebalin (1902–1963), a composer who wrote entire symphonies after he lost his capacity for language. He suffered a stroke a decade before his death, and developed 'aphasia': the inability to speak. And yet he was still musical; the symphonies he wrote in the last years of his life, in fact, are considered to be among his best. If one can have aphasia without amusia, and amusia without aphasia, is it therefore conceivable that music could be dependent upon language for its existence?

---

* I humbly suggest this could be because Sperber is amusic.

Moreover, how can music be 'useless' when it is one of the few innovations shared by every single culture worldwide? Not all – past or present – possess writing, architecture, agriculture, the number zero, or stratified societal structures. But every single one, from the Arctic to the equator, whether nomadic, agrarian or totalitarian, creates music. For such a strange behaviour it is pervasive: one of the few human behaviours that is universal. It is an indisputable component of what makes us human. We all make strange non-verbal noises together.

And what noises they are. Yet it is just the processing of sound, the product of compressed waves of air. It starts with nothing more than travelling molecules hitting our eardrums. And it can move us to tears. The more one thinks about it, the stranger this becomes.

The pleasure of sex seems sensible. How else are our bodies to be tricked into the dangerous, exhausting and expensive act of reproduction? The attraction of drugs, when placed in an evolutionary context, is also more easily explained. Considering how many millions of years our nervous systems have spent being manipulated by the chemical compounds of plants, it would be surprising if we *didn't* take drugs.

But music, upon deeper inspection, makes little sense. To create patterns of sound, to devote precious resources to the creation of electric edifices to amplify these strange noises, even to the point that we irreparably damage our tympanums. To jump, shout and sway in tandem to repetitive, bizarrely cascading sounds.

What could have inspired the thought of stretching cat gut taut for the purpose of setting it vibrating using the tail of a horse? And, even stranger, who could ever have imagined it would sound so sublime?

# SYNAPTIC SYMPHONIES

Philosophers had pondered the significance of music long before neuroscientists probed the effects of chords on the brain, for obvious reasons: music is pervasive, mysterious and power-ful. Many have achieved a decent attempt at explaining music's strange, unique qualities. Confucius made a good stab at defining its properties: 'Music produces a kind of pleasure which human nature cannot do without.' Plato considered music's power so great he would have banned most forms from his Republic. (Though, to be fair, he would have banned almost anything fun from the Republic and would have made decades of military service mandatory.)

These ideas only bring us so far.

It is in how music affects the brain that we gain enlighten-ing clues to the mystery and the majesty of organised sound. The brain is composed of 100,000,000,000 long, thin electrical

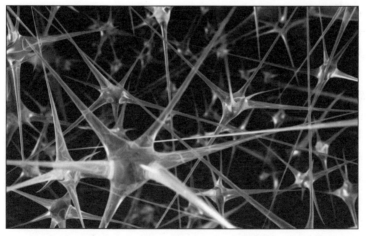

63. Any single neuron can connect to more than 1,000 other neurons: there truly are more possible connections between the cells inside your head than there are atoms in the universe.

strings called neurons – as we have noted, the brain is the most complex thing in the universe. Each of these can connect to innumerable other neurons, together weaving a dense web capable of producing thoughts, hallucinations, depressions and emotions.

There are more possible connections between the neurons in our brains than there are atoms in the entire universe. This is cliché to those familiar with it – but still a concept that even the most seasoned neuroscientist cannot fully grasp. Nobody can. The complexity of the human brain defies the capacity of the human imagination.

Wielding the modern weapons of stealth neurological espionage, scientists are able to photograph, measure, probe and play with the human brain unlike ever before. When they finally turned their sights upon music, they discovered something extraordinary. It doesn't just involve one part of the brain, or two. It incorporates regions from every corner of the brain, from the front to the back, the top to the bottom.

The evolutionary development of our brain has involved piling new structures on top of old ones: ancient structures, such as the brainstem (at the back, linking to the spinal cord), the cerebellum (which we've met before, Latin for 'little brain', the wrinkly bulb at the back of the brain), and the hippocampus (Greek for 'horse' and 'sea monster'), a pair of seahorse-shaped structures buried deep in the centre of the brain, crucial for the formation of memories.

We share these structures, which belong to a deep and ancient part of the brain known as the 'limbic system', with animals such as lizards and amphibians. Hence this cerebral constellation's nickname: 'the reptilian brain'.

Throughout evolutionary time, new structures were systematically added on top of old ones. Generally speaking, 'primitive' structures are at the back and the bottom; 'derived' or 'advanced' forms at the top and the front.

And music tickles them all.

In the cortex – the wrinkly outer surface of the brain, where we saw scientists map the clitoris, nipples, feet, hands, penis and testicles in the vertical strip of the somatosensory cortex – music stimulates several regions, such as the auditory cortex, just behind the ear, which responds to tone and timbre. Music tangos with the motor cortex, the other vertical strip that runs alongside the somatosensory cortex and is normally involved in the direction of muscular activity. The somatosensory cortex thrums when we play (or just imagine playing) music. The frontal cortex is also aroused; this sits right behind our forehead. We associate it with 'higher thoughts' and thus consider the frontal cortex to be the most 'human' part of ourselves (and thus the region we scoured with lobotomies in efforts to stop ourselves from thinking too much).

Music strums the most ancient parts of our brain, such as our familiar friend the cerebellum, one of the oldest parts of the brain, crucial for balance – and rhythm. It comprises a tenth of the brain's weight, but it contains between 50 and 80 per cent of the neurons in the brain.[46]

In the 1970s, neuroscientists discovered that not all of the projections leading from the spiral of the inner ear travel directly to the auditory cortex. Some head straight for the ancient cerebellum, where they anchor the connection between hearing and balance. Long before the harpsichord or cave lithophone existed, this served an important purpose: the 'auditory startle response' is one of the quickest reflexes our bodies boast, because we are likely to hear a predator (as we can sense it from any direction) before we can see it. The ears are hyperlinked to the reptilian brain.

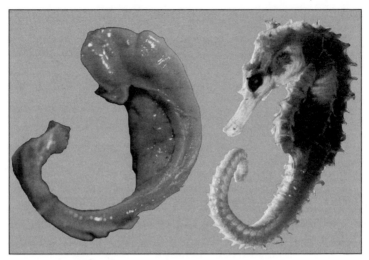

64. The hippocampus – from the Greek for 'horse' and 'sea monster' – owes its name to its resemblance to a seahorse. Small and squiggly, it is the starting point for memory formation.

The ancient hippocampi (the seahorse-shaped Greek sea monsters), positioned deep in the centre of the brain and crucial for memory formation, play their part. As do the amygdala,[47] two small clumps of cells perched just in front of the hippocampi and central to our feelings of emotion and arousal. It is instructive that the amygdala does not respond to random notes, but does respond to music.

In fact, the brainstem, the oldest part of our brain and one normally relegated to base functions such as regulating heart beat, also responds to music – but not to random noise. The cerebellum, the brainstem and the amygdala all respond during brain scans to music – but not to random notes.

Our heads react to music like an orchestra to a conductor: separate sections, all playing their parts, pipe at each other and work together to create a symphony of neural activity. This helps explain why amusia is so common, and why it comes in different forms:

malfunctions in a wide variety of brain regions can easily disrupt the complex machinery required for the perception of music.

No other human activity has been observed to stimulate as many parts of the brain; it appears as though music touches every region inside our skull. Not language, nor art, nor sport, nor mathematics can do this.

Music is an unrivalled conversation between the old and the new parts of the brain.

The more neuroscientists examine the brain, the more they discover. The discoveries continue to surprise. In 2008 Professor Aniruddh Patel at the Neurosciences Institute in San Diego found that when we listen to music, the electrical signals flowing along the neurons in our entire head pulse in synchrony in a way that has never been observed before.[48] 'That was truly a "wow" moment,' he says.[49] His choice of words is noteworthy: scientists are not encouraged to employ hyperbole.

## CHEMICAL CRESCENDOS

As they play their parts, these brain regions produce cascades of chemical flourishes: the main players in the cast of characters we have met throughout our biological surveys of sex and drugs join in.

Serotonin, the neurotransmitter we most associate with happiness and contentment, is released. Dopamine,[50] the same drug released by cocaine and heroin, flourishes. Vasopressin,[51] one of the chemical characters we met in the neuroscience of love and monogamy, join the party (especially in the realm of dance). Thrilling surges of endorphins are sprinkled on top. Music also dampens the chemical messengers of anxiety: listening to music can decrease levels of cortisol, the toxic stress hormone which debilitates and destroys, and the very reason so many people get into corrosive drugs in the first place.[52] Cortisol is poison, and music can ameliorate its effects.

Our appreciation for the chemical composition of music began with a landmark 2001 study,[53] when researchers were inspired to use 'positron emission tomography' (PET) to probe the brains of volunteers as they listened to music. They found that the more 'chills and thrills' (their words, not mine) people reported, the more activity they saw in ancient regions like the amygdala. This put us on the path to understanding how music raises goosebumps on the skin and sends shivers down our spines.

Another landmark came in 2005,[54] when Levitin found what countless scientists had suspected: music stimulates the nucleus accumbens, the 'reward' centre of our pleasure circuits, resulting in the exuberant release of our old friend dopamine. Some scientists have inferred from these overlapping neural circuits and neurochemical signals that our holy trinity was bound from the start: 'There is a link right in the brain between sex, drugs, and rock and roll,' as Levitin puts it.[55]

## HARDWIRED

As we continue to understand how music affects the brain, the more we appreciate that we are 'hardwired' for music. We are built to create it.

No group is more instructive in this regard than humans in their primacy: babies. Scientific studies of infants and music, where entire laboratories-cum-nurseries are devoted to their scrutiny, have taught us that babies are instinctively musical.

Indeed, we hardly need science to tell us that babies dig good tunes, for parents in all cultures speak to their tots in the same manner, using a lilting, melodic, repetitive, musical prosody, dubbed 'motherese'. Babies will always attune more to words spoken to them in a melodic manner than flat spoken language. The ABC song helps toddlers learn and remember the alphabet because they are hardwired for music. They will invariably pay more attention to musical sounds than disordered notes. Countless viral sensation YouTube videos of screaming babies suddenly soothed by reggae, hip hop, and even techno are testament to this.[56]

How do scientists study what kind of music non-verbal babies like? The same way mothers do: by observing their behaviour and responding to it. Scientists have meticulously catalogued what music babies express an interest in or ignore, and the kinds of emotional reactions they have to the notes tooted at them by recording if they turn towards or away from a speaker. In one early paper on the reactions of babies to music, Harvard biologists determined that 'fretting and turning away from the music source occurred more frequently during the dissonant than the consonant versions'. Conclusion: babies prefer consonant (sweet sounding) noises to dissonant (sour ones).[57]

Research on babies also revealed something few anticipated: each and every one of us is born with perfect pitch.[58] This is because the baby brain is hyperconnected: there are thousands more connections between the neurons in the brain of an infant than in the brain of an adult. It seems all babies live in a synaesthetic haze, where every smell is tinged with colour, sound is infused with colour, every smell coloured with sound. A hallucinogenic explosion where all senses blend with one another in a carnivalesque whirlwind of experience. No wonder babies perpetually look simultaneously exhilarated, overwhelmed and exhausted.

As fun as this sounds, the party can't last forever. As babies grow, these connections need to be pruned so the brain can do more with less. It needs to become more efficient so it can economically learn to make sense of the world and respond in a capable manner. But for the two heady years that the delirious trip lasts, babies possess absolute pitch. If they could speak, they would be able to name any note if played in isolation.

We lose this as we age – unless given the right stimulation. Children who grow up speaking tonal languages, such as Mandarin, where words spoken at different pitches have different meanings, are more likely to retain perfect pitch.[59] Use it, less likely to lose it.

Intriguingly, many professional musicians with perfect pitch owe their talent to synaesthesia:[60] They have retained some of the neural connections between brain regions the rest of us have lost. For them, B flat may taste like banana, or look like tinged terracotta – its distinctive colour or flavour is unmistakeable. Piano tuners frequently possess this gift, making their seemingly difficult craft effortless.

Though the rest of us lose this synaesthetic, exquisitely tuned ear, as adults our brains are far more primed for music than we might realise.

Our capacity to remember tunes, and recognise songs even from a split second of a recording, is extraordinary. Pick some of your favourite tunes sometime, and see if your friends can recognise the track from a single note. It's fun. I recommend the cymbal crash at the start of Radiohead's 'High and Dry', the first piano note at the start of 'Fairytale of New York' by the Pogues (or the harmonica at the start of 'Dirty Old Town' for that matter), and the guitar strum which opens 'Devil's Haircut' by Beck. Familiar audiences nail them, every time. The Radiohead cymbal crash is particularly impressive: it derives from a standard drum kit, and yet the combination of that crash in that studio on that recording refined with that production has been so hardwired into our melodic memories crowds recognise it instantly.

Our brains are hardwired to remember melodies, and we have used this to our advantage for thousands of years. We remember words set to music far easier than straight spoken prose. Nursery rhymes are melodic because children find them easier to remember. Epic ballads, sung and not spoken, were passed down for thousands of years without written inscription. People who work with the elderly and have witnessed the ravages of Alzheimer's will attest that often music is the last thing to go. Even when names, identities and memories are forgotten, the songs of our youth remain.

Though many of us may doubt our ability to keep a beat, even the least coordinated of us are primed for percussion: the capacity of our brains to perceive a beat and maintain a steady rhythm is considered by some scientists to be a unique quality of the human mind. Students of physics, take note: Galileo recorded the speed of gravity by employing music. He felt the clockwork mechanics of Renaissance Italy were insufficient for his purposes: so as he dropped objects and measured the speed of their descent, he sang to himself to measure time. His calculations ... they weren't half bad.

## NEURAL NUTRITION

Everything we now know about how music affects the brain can be put into practice in ways that are transformative. Children who are given music lessons will find that their achievement in other subjects – from mathematics to sport – will improve. IQ, rough an estimator of intelligence as it may be, increases.[61] Adults and children alike will perform better on tests, puzzles and problem-solving exercises. Researchers have actually observed changes in the brains of children given music lessons compared to those left wanting: a 2009 trial in Boston gave piano lessons to 15 children, and none to an equal number. Result: measurable changes in the size and shape of the regions of the brain that play a part in auditory and motor skills.[62]

Much is made of the so-called 'Mozart effect', the perceived benefits of listening to classical music, and Mozart in particular. In fact, only *one* study has ever demonstrated that Mozart is good for the brain: a small investigation conducted in 1993.[63] People were given ten minutes of Mozart, a relaxation tape, or silence, and then asked to complete a fairly tricky exercise: the paper-folding and cutting task. This involves imagining a piece of paper folded several times, several corners cut, and then correctly identifying the jagged shape once unfolded (a bit like making paper snowflakes).

Those who were played Mozart did marginally better. The idea of classical music as neural nutrition was so appealing, the market is today inundated with baby toys, study aids, prenatal bump speakers, and other 'neuro' products that peddle classical music as an educational elixir.

But more recent studies have revealed that the same benefits can be induced by listening to the music that one just happens to like[64] – no matter what it is – leading to the effect being cheekily redubbed 'The Blur effect'. This therefore says little about the impacts of particular musical styles on the human brain and far more about the cultural backgrounds of the scientists involved. Old, white and male, as the saying goes. With changes in society come changes in discovery.

This implies, therefore, that the scientists who conducted the

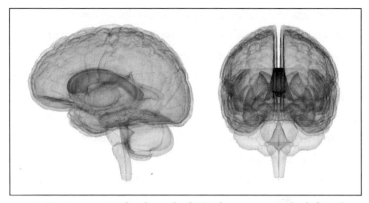

65. Music seems to be the only thing that connects the left and right hemispheres of the brain together, through the corpus callosum, a structure that lets the two halves 'talk' to each other.

initial study that purported to demonstrate intellectual enhancement due to exposure to the music of Mozart could have found a correlation between the music of any composer and any population, so long as it was music that people liked. If the investigators had surveyed enthusiastic heavy metal fans, would they have discovered the 'Judas Priest effect'?

All music that pleases is neurologically beneficial. But more significant, music changes the brain in anatomical ways that are actually visible. 'Anatomists would be hard put to identify the brain of an artist, writer or mathematician – but they could recognise the brain of a musican without a moment's hesitation,' as doctor and famed populariser of biomedical science Oliver Sacks puts it.[65]

The grey matter – the layer of neural endings covering the outer surface of the brain – thickens.[66] The auditory cortex, that horizontal strip on the outer surface of the brain behind the ear, grows bigger. The cerebellums – our 'little brain' based at the rear, crucial for balance and movement – have also been shown to be larger in pianists.[67] This might come as no surprise, considering the coordination involved in manipulating ten fingers at once to produce the effect of an entire orchestra.

There is one particularly pronounced and unique change that comes with musical training: the corpus callosum, the network of neurons that connects the left side of the brain to the right, becomes thicker.[68] In most people, the left hemisphere appears to process rhythm while the right coordinates melody.

We still don't know what a thickening in the corpus callosum means, but it seems that only music creates this unison. Beliefs that the left side of the brain is 'rational' and the right side 'artistic' – or that the left and right side are fundamentally different at all – are a matter of scientific controversy. Nonetheless, if music is a unique conversation between the old and the new parts of the brain, it is also an unrivalled link between the two sides of ourselves.

There is no better way to summarise everything that we know about music than this: music is good for the brain. Even if it's Judas Priest.

## MEDICINAL MUSIC

If 'retail therapy' is one of the ugliest possible pairings in the English language – mindless material consumption as transient psychological consolation – then 'music therapy' has to be one of the loveliest linguistic duets imaginable. Music as more than mere entertainment: music as medicine.

Melodies and rhythms can be employed as therapeutic interventions for a multitude of conditions: depression,[69] stress and anxiety can all be alleviated by listening to or making music. We do this daily to soothe ourselves during unpleasant moments, even if unconsciously (observe how many people on the street are habitually plugged into their iPods). Carefully designed scientific studies have found that music can be employed therapeutically, often with remarkable results. Children suffering the ordeal of a dental visit require less anaesthetic if played music.[70] Autistic youngsters can be coaxed into social interactions with music.[71] Students with learning difficulties can improve their performance in mathematics and linguistics with music.[72]

Music therapist Helen Mottram was classically trained in the euphonium (a miniature spunky tuba), and retrained as a music therapist. She now specialises in work with children with learning difficulties and the scars left by abuse or neglect. 'After just one day of an introductory course to music therapy during my undergrad, I came away and knew this was what I wanted to do for the rest of my life,' she says.[73]

'It hasn't just changed the way I relate to children. It has changed me as a musician in every way possible and the way I think about music: the role it has in daily life, and what music itself ultimately is,' she says. 'When I was a student, music was all about recreating the symbols on the page as accurately as you possibly could. And that is what music was to me. Over the years, I learned that music can be anything – from the scratching of pencils to a banging spoon. What makes it "music" is the meaning it has for you.'

Even in our later years, music can still be used medicinally to help us recover things that seemed lost forever. Stroke victims who develop aphasia can learn to speak again through 'melodic intonation therapy',[74] just as children learn the alphabet through song: through music, we can restore language. People suffering from Alzheimer's, as many of us have seen ourselves, will remember the songs of their youth far better than yesterday's news. This can be put to therapeutic use to help them regain speech and memory skills.[75] The potential for using melody to replicate the way we learn to speak as children is obvious, yet the first scientific study on the therapy's potential was only published in 1973.[76]

But here's something I find even more arresting: of 54 stroke victims who were given either audio books or musical tracks to listen to as part of their medical recovery regimen, those given music improved in their linguistic capacities to a greater degree than those given recorded words.[77] Music aided linguistic improvement better than linguistic treatment itself.

Muscularly, music can work wonders: those partially paralysed following stroke can regain the use of their limbs through 'gait therapy', pacing their stride with drums, rhythms and percussion to relearn how to walk.[78] Therapeutic effects can also be physically transformative for those hindered from birth: many individuals with Tourette's syndrome find that playing and hearing music alleviates twitches. 'The ticks will go away as soon as I start to play,' says Nick Van Bloss, pianist and former lead musician in the English Chamber Orchestra, told the *Guardian* newspaper in 2010.[79] People living with Tourette's have even been known to form drumming circles, so effective is the communal activity.

And one impact of music therapy seems more remarkable than any other: it can speed the healing of hearts, lungs, and other vital signs of premature babies in neonatal intensive care units.[80] Placebo effect? Perhaps not.

## SONIC BONDS

Any explanation for the importance of music has to take into account that it is a social activity. Our elevation of music into a staged performance is a fairly recent innovation. For thousands of years – and still in many places around the world – it is a communal phenomenon. It is not a spectator activity, practised by the few for the enjoyment of the masses. It is something we do together. 'Music is a form of social negotiation via sound,' as Aniruddh Patel puts it.[81]

In 2009 Austrian researchers built on Patel's discovery – that the neurons of the brain synchronise their firings when we listen to music – by hooking EEG kits up to the scalps of duetting guitarists. They found that their patterns of firing in the somatosensory cortex[82] (the vertical stripe along the top of the head which we've met throughout this book, the map of feeling) will begin to harmonise their electrical buzzing. What does this mean? It's not yet clear – but it is at least obvious that the social links we create when we make music together create biological symphonies we are just beginning to understand.

It is in fact the social qualities of music that – from the standpoint of cognitive science – could be much better understood. 'I think we need to understand music better in a social context, with other people, playing or dancing, rather than listening – this is poorly understood at the moment,' chimes in amusia specialist Dr Peretz. 'Music is ultimately a social activity.'[83]

Musicologists and philosophers have generally noted this, and recent scientific investigations have confirmed that making music together binds us in a chemical fashion. Professor Robin Dunbar at the University of Oxford has investigated drumming circles and choral singing, and his work indicates that producing music together leads to surges in 'endorphins'[84] – that feeling of being 'high'. More instructive: singing in choirs produces oxytocin,[85] the 'cuddle chemical', essential for bonding and attunement, also produced by orgasms and breastfeeding.

For many musicians, the result is 'magic' – or 'jam magic' to be precise. Guitarists will discover that difficult fingerings and chord progressions become easier; trumpeters may suddenly reach high notes which before they may have struggled to attain.

There is one finding that is perhaps more instructive – and, frankly, more remarkable – than any other. When asked to keep a beat, many of us consider our timekeeping skills to be less than adequate. Students often use metronomes to keep time: in the past, gear-driven ones, and now, electronic. But here's the extraordinary thing: in laboratory tests, people synchronise beats and keep accurate time better with other humans than they do with machines.

The ubiquitous capacity of our species to synchronise beats in groups has been proposed as a plausible explanation for how pre-industrial societies constructed monumental works of architecture. Rather than alien intervention or some form of long-lost magic, the pure combination of our innate rhythmic capacities with collective muscle power (albeit sometimes that of hundreds of slaves) allowed us to haul the giant stones of Stonehenge or the building blocks of Mayan temples. 'Heave-ho, heave-ho.' If we didn't have a sense of rhythm, we may never have built the pyramids.

Dr Carl Lipo, an archaeologist at the California State University, had a hunch that rhythmic coordination erected the statues of Easter Island. The standard narrative holds that the enigmatic heads travelled to the top of their hill on their backs, rolled on logs by crazed islanders who chopped down the trees of the remote island in a delusional quest to erect statues of worship. Anthropologist Jared Diamond of the University of California upheld this as a prime example of humanity's innate tendency to destroy its own home in his 2005 book *Collapse.*

But Dr Lipo felt there was something amiss with the standard

66. Heave ho! Carl Lipo argues that the heads of Easter
Island made their way up the hills by walking rather than
rolling. Our ability to keep time together was crucial.

narrative. One, the road from the quarry leading up to the hilltop
is littered with broken statues lying face down. Two, the trees of
the island are monocots (a family of plants that includes grass,
sugarcane, palm trees and bamboo): structurally unsound to
support the statues. And three, the centre of balance of the statues
lies at the front of the belly. Put everything together, and it suggests that the statues didn't roll up to the hilltop: they walked.

In *The Statues That Walked* (2011) Dr Lipo argues that Easter
Islanders did not chop down their forests in a crazed act of self-
sabotage. Instead, he asserts that they used ropes to rock the
statues this way and that, yanking them up the hill by creating
motion through gait. 'They weren't crazy – they were clever,' says
Dr Lipo.[86] 'They didn't destroy their environment because they
were primitive. They did this because it was fun. Think about it:
they lived on a remote island on the Pacific. There wasn't much to
do. This was something they could do on a weekend.'

His contemporaries sniffed at the idea. Then American film
station NOVA gave him a wad of cash to just try it out. Forget

the equations, and just do it. So they did. And it worked. They published their study in 2012.[87] And you can see the video online: 12 people, a few ropes, a 10-metre high replica statue, the power of human rhythmic coordination. And a bit of monetary muscle.

## MUSICAL MONKEYS

It comes as no surprise, then, that some of the oldest known human creations are musical instruments. Recall that the oldest ever recovered could be a flute. It is clear that music has been with our species for a very long time. Could it have even been with us before we could communicate in any other way via sound? In other words, does music predate language? Or, perhaps, were they once one and the same?

Darwin was one of the first to propose that music and language were originally blended in a melodic 'protolanguage'. Jean-Jacques Rousseau also believed that language and music were originally mashed together, remnants of which are retained in 'motherese'.

As mentioned, anthropologist Steven Mithen feels that music is ancient and was present in a wide swathe of hominids, from Neanderthals to ourselves. How, then, did humans learn to speak? Mithen expands on the 'musical protolanguage' idea, and suggests that early *Homo sapiens* developed a dancing, gesticulating, sing-song form of communication he calls 'Hmmmm' (for holistic, manipulative, multimodal, musical and mimetic).[88]

Mithen's idea strikes at one of the most crucial characteristics of music: dance. The coupling of sonic experiments with muscular movements. Indeed, in many cultures 'music' and 'dance' are not two words, but one.

True, not everyone loves to dance – many find the experience downright humiliating – but the irresistible drive to move in time to music is found in all cultures, and takes a bewildering, spinning, dizzying, pirouetting array of forms.

Is dancing a recent innovation? Or is it tied to the evolution of music? The sacculus, that ancient fishy sac in our vestibular system and a key player in our motor reflexes and our balance, triggers electrical signals in the neck when stimulated with loud music. There are direct links from the ear to the motion-sensitive cerebellum and to the sex-fiend drug-dealing nucleus accumbens.

Is the urge to get into the groove embedded in our anatomy, programmed by evolution?

There have been very few scientific studies on the nature of dance, and most make loose psychological correlations, such as one that linked 'personality traits' to dancing style[89]. Another claimed[90] that women preferred the moves of men who had been exposed to more testosterone *in utero*, as revealed by lower index to ring finger ratios (2D : 4D, which we met in the survey of science and sex). Finger ratios, as noted, are supposedly biological markers for levels of 'male' hormone exposure in the womb. Blogs predictably headline stories with charmingly reductionist headlines such as 'Men's dancing style determined in the womb'.[91] The publication of this study proves once again that scientists are people, people are flawed, and flawed people can publish prejudiced findings with what they purport to be intellectual authority.

Yet another survey claims middle-aged men dance in a cringe-worthy fashion ('dad dancing') because evolution is driving them to repel young fertile women into trysts with strapping, more virile lads.[92] Make of that survey – 'Dance confidence, age and gender', based on surveys with 14,000 people – what you will.

As we have noted, studies of the mind are prone to statistical skewing, because the mind is complicated, everybody is different, and extracting clear conclusions from mountains of complex data is maddeningly difficult. Though genetic discoveries can thus seem more concrete and 'real' by comparison, the impacts of single genes – and the myriad effects they produce throughout the entire body – are still poorly understood. So the behavioural implications of genetics, though fascinating and impressive, should be taken with the same iceberg of salt (as we first discovered in our survey of the history of homosexuality in science).

Nonetheless, some recent studies have found intriguing links between genes and dancing prowess. A comparison[93] of pro and semi-pro Finnish dancers, their relatives, and other physically skilled Finns (athletes and suchlike) revealed that dancers shared a gene. Humbly named 'AVPR1a and SLC6A4', this gene translated to higher levels of serotonin – the neurotransmitter LSD and psilocybin masquerade as – and the hormone vasopressin, which we met in Part 1, linked to sociability, monogamy and bonding. The Finnish authors think that the link between the neurotransmitters and hormones and dancing talent has more to do with innate sociability rather than muscular or athletic aptitude. Another study by the same Finns found that genes coding for the amorous vasopressin are found more commonly in professionals boasting 'musical aptitude'.[94] The scientific study of dance is young, and the possibilities for examining the evolutionary roots of our need to bounce in time to a tune are boundless.

An idea has been kicking around my head for about six years – since the very first time I gave a short lecture on the neuroscience of music in a field with Guerilla Science at a music festival in 2008.

My best friend, a shabby musician, asked me: why did we

come down from the trees and stand on two legs? Did it have anything to do with music?

An idea came to me then, and it hasn't left me since.

Maybe we stood up because we wanted to dance?

Bipedalism remains a defining feature of our species, and numerous theories have been proffered as to what brought our ancestors down from the trees and coaxed us to stand upright. The 'aquatic ape hypothesis' – which posits that we lost our body hair and gained our upright posture (and fatty frame) as a result of time spent wading in warm, wet shallows – is highly appealing. But there is poor evidence for it in the archaeological record. The scientific basis for the importance of running in the evolution of our species is pretty rock solid.[95] But obviously we had to walk before we could run. Why exactly we stood upright in the first place remains a matter of debate.

Could it be that the desire to dance drove our transition from awkward ambling apes into the cavorting, dancing athletes we are today?

Now, I am a science writer not a researcher. I have no data to justify my idea.

I'm just putting the thought out there. If a scientific researcher with the credentials, resources and time has the inclination to test this and provide weight for the idea, I'll be pleased as punch.


## DUETS WITH NIGHTINGALES

If music predates language, and perhaps has been with us since our earliest days, this leads to the obvious question: Did our musical sprouts spring from even older roots? Do animals make music?

Undoubtedly, the world is bursting with zoological noises which we can call beautiful: chirping tree frogs, singing whales, and, of course, all the thousands of varieties of birdsong certainly sound like music. Even some species of fish 'sing', thudding their swim bladders with their stomach muscles and producing a noise like a didgeridoo.

Scientists only discovered that whales produce complex soaring calls lasting up to 21 hours a time in the 1960s. Long-haired beatniks wasted no time heading out to the beach with flutes and clarinets to chime in. Whale biologists, alas, had to point out that they probably weren't duetting, and were most likely just annoying the animals. Funny as these interspecies aquatic 'jam sessions' might seem, decades earlier, the world's very first outdoor radio recording in 1924 showcased cellist Beatrice Harrison performing spontaneously with a nightingale,[96] which had perched unheeded alongside her and sang as she played.

But if we know that music is something our minds create – how can we possibly claim to know what nightingales and whales hear? Do they experience disjointed noise when they hear a piano? And what do they sound like to each other?

Intriguingly, researchers have found that animals can be trained to distinguish between different styles of music – even carp[97] can tell the difference between the blues and classical scores. But do they like it? Are they simply demarcating the difference in tempo or pitch? Do they actually perceive anything of value in the sound waves?

Any clue to the roots of human music should be sought in our closest relatives: monkeys, apes, and other primates. Which invites the question: What kind of music do monkeys like? (Surely the answer must be jungle.)

Cognitive neuroscientist (and former club DJ) Josh McDermott spent more than a year trying to answer this question with common marmosets and tamarins. He would place the primates into a V-shaped maze, in which a speaker at one end would play one genre of music, and a speaker in the other something different.

The monkeys could choose which music they preferred by going into that arm of the V, just like addled ravers in a multi-room rave. McDermott gave the monkeys a variety of tunes to choose from: Mozart, lullabies, techno and silence.

He found (perhaps to his profound disappointment) that monkeys would always opt for silence.[98] His tamarins and monkeys preferred music of a slow tempo to fast-paced dance music, but invariably their favourite tune was silence. It seems the small primates were nothing but irritated by the sonic combinations of *Homo sapiens*.*

Among our kin, we are alone: we are the musical monkey.

Some zoologists conjecture that it is possible that other species of animal who sing to communicate, find partners, guard territory and fight might 'hear' music in some way that is similar to us. If so, it would only be those species who display 'vocal learning': birds, whales and seals (who have to learn their songs through practice) may be 'musical'.

Dr Patel, though he admits animals have sonic skills, once asserted: Other creatures may be melodic, but humans are the only species that can keep a beat. Other animals may have tuneful tones in their calls, complex structures in their sonic communication – but only humans can synchronise their motions with a regular beat.

---

* Although some larger primates might be more musical than McDermott's marmosets. Bonobos at the Great Ape Trust trained in sign language have been known to enjoy playing with tambourines, maracas, drums and xylophones, and have been treated to jam sessions with Peter Gabriel and Paul McCartney. Whether or not they are enjoying 'music' – or simply revelling in making noise – remains to be determined.

67. In 1924 a nightingale joined the world's first outdoor radio recording when it sat alongside cellist Beatrice Harrison and performed an impromptu duet.

68. What kind of music do monkeys like? Neuroscientist and former DJ Josh McDermott spent more than a year playing dozens of different genres to common marmosets.

69. Snowball the crested cockatoo demonstrated in the lab that it could stomp its foot in time to music: a scientific first for an animal.

His offhand statement set off a flurry of infuriated emails from incensed internet users, adamant that their cat was an avid Bon Jovi zealot, their dog a die-hard Bob Marley fan.

Fine, he said, in the spirit of scientific inquiry. If you have a pet in the San Diego area, and you think it can keep a beat, bring it to my lab and I will study it.

And so he did. Snowball the crested cockatoo came to Dr Patel's lab in 2008, and lo and behold, it could indeed keep a beat.[99]

Videos of Snowball stamping its foot in perfect time to the Backstreet Boys and Michael Jackson have been viewed on YouTube millions of times – Snowball getting down to 'Another One Bites the Dust' by Queen has garnered nearly 5 million hits.

Then came Ronan the seal of the Long Marine Lab in California, whose trainers taught it to bop its head in time to disco tracks.[100] These animals, however, live with humans, so are not likely to be representative of their species.

# INSIGHT AND INSPIRATION

If science is perceived to be an uninspired reduction of life's sublime mysteries into grim mechanical mundanities, few subjects could refute this better than the neuroscientific study of music. It has shown us that nothing affects our brains like music. Whenever we hear its chords we experience a symphony of neural tremblings, resulting in a firework display of exhilarating chemicals, sending shivers down our spines, raising goosebumps on our flesh, and kindling a burning in our hearts. It is ancient, it is hardwired, and one of the key ingredients that made us human. It took 3 billion years to appear.

We don't need it to survive, but evolution gave it to us nonetheless. It is a gift.

But equally instructive is what music has to teach us about science. Mainstream proponents of the view that we have a responsibility to 'educate the public about science' and so forth will expound on music as one of the clearest examples of the ways that science can enlighten, explain and inspire in ways that philosophy, theology and the arts cannot. No ancient thinker could have foreseen that within our heads billions of tiny strings are strung, all reverberating in unison when we hear music. This is true.

But neuroscientists would never have come to this conclusion had it not been for the fact that so many of them had spent decades training, practising and performing music, gaining an appreciation of its importance to a depth they never could have gleaned from a textbook. Artistic experience inspired scientific insights.

More importantly, if young scientists had held too deep a reverence for the ideas of their superiors, it is unlikely we would have grown to understand either music or the brain. Rebellious research inspired by intuition, experience, and, above all, passion is not just helpful, but crucial to intellectual progress.

# CONCLUSION

What unites the unholy trinity of sex, drugs and rock 'n' roll?

All are ancient features of our biology, each produces cascades of delicious neurotransmitters, and every one has been frowned upon or restricted by religious and political authorities. Even in the past few years Islamic zealots have tried to outlaw music in Mali, a nation that has one of the world's richest and most colourful musical lineages.

Sex, drugs and music are three vastly different things. Sex, though sometimes a source of misery as well as joy, is essential for the creation of life. Drugs, on the other hand, are indisputably unnecessary for the experience of a happy and healthy life and the cause of intense suffering and physical deterioration worldwide. Music, a world apart from sex and drug use, is probably the strangest thing we humans do. Of our hedonistic triad, it is the one that is most uniquely human, at the same time completely gratuitous and yet one of our species' greatest achievements. It is both superfluous and spectacular.

What, then, binds these three very different things together? Surely something must, as the phrase 'sex, drugs and rock 'n' roll' is embedded in our lexicon and our collective consciousness. It's tempting to suggest that there is an exuberance that can be found in the enjoyment of all three of these things that other human pursuits just cannot match. This might just be my perspective, but I've never seen grown men moved to tears at an art gallery, sports match or political debate as I've witnessed in the opera house, in the field of a music festival, or on the quaking floor of a nightclub. What could draw 500,000 dirty sodden teens to sleep rough for four days in a cramped field in upstate New York but the power of sex, drugs and music?

There is a primal power in these three things that touches us on a deeply emotional level. And, though not all of the best things in life are entirely free, these are about as free as it gets. They are evolution's gifts.

Some scientists have dubbed the nucleus accumbens the 'sex, drugs and rock 'n' roll' centre of the brain, as all three indulgences can trigger the release of our tasty neurotransmitter dopamine in this region of the brain. But this is an oversimplification: gambling, overeating, laughing and learning (an underappreciated indulgence) can all trigger the release of dopamine from this tiny corner of the brain.

Is there any biological root that unites sex, drugs and music to the exclusion of all other hedonistic joys? If there is, science hasn't revealed it yet. And there probably isn't one.

But what science has taught us about all three things is that, contrary to being base or animalistic pursuits, they all comprise important facets of the human condition. Over evolutionary time they have all become some of the most colourful and dramatic defining characteristics of the human species. Other animals may rut, gorge on narcotic plants, and make exuberant noise – but we have taken all three to the next level. No animal climaxes with the same ferocity, takes (or makes) drugs with the same single-minded devotion, or creates such an incredible range of sounds with the intent of inspiring such a wide range of emotions.

Our bodies are hardwired to experience orgasms more

powerful than seemingly any other animal on the face of the earth, and we possess a range of physical quirks – from alluring breasts to ostentatious penises and trigger-happy clitorises – that all seem to have predestined us to a lifetime of sexual longing, pursuing and indulging.

Our brains are wired to twig to a vast range of chemical stimulants, and we almost certainly do a wider range of drugs than any organism that has ever lived. Plants have spent millions of years committing botanical burglary, lock-picking the genomes of animals and insects across the evolutionary tree, and they have had more success seducing our supposedly sophisticated brains than any other form of life on earth. Clever we may be, but easily manipulated as well. We are primed to go weak at the knees for botanical and fungal trickery.

Our vocal cords and our ears are exquisitely primed to create and attune to noise, and synchronised performances between our muscular reflexes and the timekeeping centres of our brain have gifted our species with a sense of rhythm that is unsurpassed on this planet. Music, in fact, strokes more parts of our brain than anything else we encounter or perform. And as we recently discovered, alone among our earthly delights, it creates harmonised symphonies of electric neural singing.

Moreover, we have used our opposable thumbs, puzzle-solving brains, and inventive imaginations to find new ways to enjoy sex without conception, create new chemicals to addle the mind, and manufacture sounds stranger and louder than any made by any living creature before us.

Is it possible to enjoy life without all three of these things? Certainly. Thousands of years of monastic traditions, the life-affirming joy recovered addicts swear is better than any high, and the happy and otherwise normal lives of amusics all testify that, of course, one can live without sex, drugs or music. But is such a life preferable? When our biology has programmed us over 3 billion years to take off our clothes, down another shot, and turn up the volume, it's tempting to think that 'hedonistic' impulses simply need a rebrand. Perhaps 'sinful' activities aren't entirely deserving of the label 'bad'.

'Most fun things in life are bad for you, but nothing will kill you faster than having no fun at all,' as my mother says.

Perhaps we label our 'hedonistic' impulses as selfish and sinful because they are so powerful, and we are simply scared of our tendency to lose ourselves – and our minds – in their pursuit. But rather than rejecting these pursuits as primitive, animalistic or degenerate, perhaps we might do better by gaining a deeper understanding of our deepest desires.

As we have seen over these pages, the end result of exploring our 'base' pursuits is a stunning array of profound intellectual insights. Brave scientists who challenged the accepted beliefs of their time have revealed how unique our sexuality is, elucidating the unknown complexities of our reproductive quirks, mapping the broad spectrum of our behaviours and desires, and time and time again fighting for tolerance in the face of prejudice. Drugs, though 'bad' for the brain, have led to profound insights into the mechanics of the brain, helping us to discover the chemical messengers of the nervous system, invent the gift of anaesthesia, and treat some of our most intractable miseries with their potent chemical trickery. And music, denigrated for decades by the world's most esteemed neuroscientists, has revealed more than perhaps anything we do that everything we experience is made by our minds: our brains make our world.

Dangerous and occasionally poisonous they may be, but none of our hedonistic desires should ever be deemed trite. They are universal, they are powerful, and they have been woven into the fabric of our species over 3 billion years of evolution.

The more we observe conventions which consider sex sinful, drugs shameful, and music inconsequential, the more we inhibit ourselves from our ultimate goal: understanding ourselves.

# EPILOGUE

It began over a mutual suspicion of Richard Dawkins, on a stack of hay bales, under the stars and surrounded by circus tents. With a techno soundtrack.

'I think he does science a disservice by reinforcing the stereotype of scientists as arrogant, snooty, misanthropic egomaniacs who just want to prove that they are smarter than everyone else,' I said.

'He just rails at one point over and over, and there are no nuances to his thinking or subtlety to his delivery,' she responded.

Jen Wong and I met for the first time at the Secret Garden Party – England's retort to Burning Man – where we had been brought together to host the science tent, dubbed Guerilla Science Camp. She worked at London's Science Museum, where she had cultivated millions of brilliant ideas that could never see the light of day within a bureaucratic institution. I came from a family of rock promoters and hippies. The click was instant.

Over six years Jen, myself, co-founder Mark Rosin, and later recruits Jenny Jopson, Louis Buckley and Olivia Koski – all scientists by training but even producers by profession – have taken Guerilla Science and doubled it in size yearly, growing from a tiny outfit hosting lectures by scientists in wee tents at giant music festivals to a multinational theatrical company hosting banquets of filthy food within Victorian sewage cathedrals, psychiatric treatments inside impenetrable white cubes amid rivers of mud, and – one day – giant invisible mazes hovering above Rio.

Every creative team has different players, and everyone has their role: The Bureaucrat. The Fixer. The Maker. The Documenter. The Actor.

But every outfit needs A Captain. A Director. A Mastermind.

It was Jen who saw from the start that our purpose wasn't simply to celebrate science with exuberance. Or (heaven forfend) 'enhance the public understanding of science' as so many in science communication see it.

Jen saw the big picture. She saw from the start that our aim was to bring the sciences and the arts back together in explosive fashion, simultaneously futuristic and historic: a return to Renaissance values using the tools of modernity.

It was Jen who forged a relationship between performance stress researchers and beatboxers. Who thought nothing could fit within the science tent better than shadow puppets. Who thought of pairing a rapper with a physicist to create a viral video.

It was Jen who said, 'I don't just want to give people facts – I want to ask them questions.' Who had the idea to make origami unicorn giveaways for a theatrical event celebrating the film *Blade Runner*, with small slips of paper prompting: 'How can you be sure this memory is yours?'

I'll never forget the text message: 'Am in Soho. Had a brainwave: let's make the origami out of Chinese newspaper. I'm carrying a stack of them.'

It was Jen who had the wherewithal to call up a flutist friend to perform with our Reubens tube the night we debuted the flaming device at a cockney working men's club. Who had the charm to borrow nineteenth-century spectacles for an event that would be packed with a thousand drunk, whimsically disobedient punters?

Who thought, why don't we build a giant synaesthetic brain that makes weird noises when presented with different colours, and turn a lecture about the neuroscience of sensory blending into a game that forces costumed revellers to dance inside a square of hay bales?

Who tirelessly made panna cotta brains (a gruesomely laborious process, if you've ever tried your hand at custard) every single evening for a fortnight as we lived and worked inside the recreated medical prison of *One Flew Over the Cuckoo's Nest*?

When we were offered the chance to host a science event within a Valentine's ball at Battersea Power Station, I asked: 'How will we compete with the trapeze and the hot tubs?' She said instantly 'I think we should do a life drawing class pairing a body paint artist and a neuroscientist.' It worked.

Who – without a moment's hesitation – signed us all up to perform an unparalleled act of scientific theatre in the heart of Glastonbury's 'naughty corner' (and for Glastonbury to even *have* a naughty corner is saying something)? Who put her ego aside and performed 'wellie runs' for two hours, carrying muddy boots backstage from one end of a giant white theatrical cube to the other?

While making a documentary about Glastonbury at the end of the exhausting weekend, a film crew asked: 'What makes Guerilla Science different from the other folks who put Brian Cox on stage inside a big comfy tent with chairs and a Power Point projector?' Tired – exhausted – all I could say was: 'None of them have the balls. They couldn't do it even if they tried.'

Jen has brought us to places I never thought possible. I grew up in the music industry, playing with drum kits during the setup at gigs, hanging backstage with circus sideshow freaks, and then bartending at a rock venue for four years.

As I said to Jen: 'I have worked with strippers, endocrine

physiologists, contortionists, anuran developmental biologists, gay leather daddies, microbial engineers, and body modification artists. But I have never worked alongside midgets smashing a classic car with baseball bats.'

But from the beginning, when the idea was a mere nugget, it was Jen who invited us over to her flat, made a vat of stew infused with carrot and cumin, and got right down to business. Right we've done the lectures in a tent: what's next?

Over four years there have been many moments of shared inspiration, a number of disagreements, a few fights and tears. Her drive to do something new has nearly bought us both nervous breakdowns.

But the fact is: none of this would be possible without her. Every ship needs a captain, every dream a mastermind.

Without Jen Wong, this book would never have been possible.

# FURTHER READING

## Sex

Dabhoiwala, Faramerz. 2012. *The Origins of Sex: A History of the First Sexual Revolution*. Penguin.

Hamilton, David. 1986. *The Monkey Gland Affair*. Chatto & Windus.

Judson, Olivia. 2003. *Dr Tatiana's Sex Advice to All Creation: Definitive Guide to the Evolutionary Biology of Sex*. Vintage.

Kringelbach, Morten. 2008. *The Pleasure Center: Trust Your Animal Instincts*. OUP USA.

Ladas, A. K.; Whipple, B.; Perry, J. D. 1982. *The G Spot: And Other Discoveries about Human Sexuality*. New York: Holt, Rinehart, and Winston.

Magniati, Brooke. 2012. *The Sex Myth: Why Everything We're Told Is Wrong*. W&N.

Maines, Rachel. 2001. *The Technology of Orgasm: 'Hysteria,' the Vibrator, and Women's Sexual Satisfaction*. Johns Hopkins University Press.

Moalem, Sharon. 2009. *How Sex Works: Why We Look, Smell, Taste, Feel, and Act the Way We Do*. Harper Collins.

Parry, Vivienne. 2009. *The Truth About Hormones*. Atlantic Books.

Pisani, Elizabeth. 2008. *The Wisdom of Whores: Bureaucrats, Brothels and the Business of AIDS*. Granta Books.

Roach, Mary. 2009. *Bonk: The Curious Coupling of Sex and Science*. Canongate Books.

Ryan, Christopher, and Jetha, Cacilda. 2010. *Sex at Dawn: How We Mate, Why We Stray, and What It Means for Modern Relationships*. Harper Collins.

**Drugs**

Brooks, Michael. 2011. *Free Radicals*. Profile Books.

Escobar, Roberto. 2010. *Escobar*. Hodder Paperbacks.

Goldacre, Ben. 2012. *Bad Pharma*. Fourth Estate.

Hoffman, Albert. 1979. *LSD: My Problem Child*. Multi-Disciplinary Association for Psychedelic Studies.

Holmes, Richard. 2008. *The Age of Wonder*. Harper Press.

Huxley, Aldous. 1954. *The Doors of Perception*. Random House.

Jay, Mike. 2010. *High Society*. Thames & Hudson.

Jay, Mike. 2000. *Emperors of Dreams*. Dedalus.

Kaiser, David. 2011. *How the Hippies Saved Physics*. Norton.

Kringelbach, Morten. 2009. *The Pleasure Center*. Oxford University Press.

Leary, Timothy, Ralph Metzner and Richard Alpert. *The Psychedelic Experience*. 1964. Reprinted as a Penguin Modern Classic.

Lilly, John. *The Scientist*. 1998. Ronin Publishing.

McKenna, Terrence. 1993. *True Hallucinations*. Harper Collins.

Nutt, David. 2012. *Drugs Without the Hot Air*. UIT Cambridge.

Roberts, Andy. 2012. *Albion Dreaming*. Marshall Cavendish Editions.

Shulgin, Alexander, and Ann Shulgin. 1991. *PiHKAL*: a Chemical Love Story. Transform Press.

Siegel, Ronald. 1989. *Intoxication*. Pocket Books.

Strassman, Rick. 2001. *The Spirit Molecule*. Park Street Press.

Wolfe, Tom. 1968. *The Electric Kool Aid Acid Test*. 1968. Black Swan.


**Rock 'n' Roll**

Ball, Philip. 2011. *The Music Instinct*. London, Vintage.

Byrne, David. 2012. *How Music Works*. Canongate Books.

Collin, Matthew. 1998. *Altered State: The Story of Ecstasy Culture and Acid House*. Serpent's Tail.

Cunningham, Mark. 1999. *Good Vibrations: A History of Record Production*. Sanctuary Publishing.

Levitin, Daniel. 2006. *This Is Your Brain on Music: The Science of a Human Obsession*. Penguin USA.

Mithen, Steven. 2005. *The Singing Neanderthals: The Origins of Music, Language, Mind and Body*. London, Weidenfeld and Nicolson.

Nyman, Michael. 1999. *Experimental Music: Cage and Beyond*. Cambridge University Press.

Sacks, Oliver. 2007. *Musicophilia: Tales of Music and the Brain*. New York, Vintage.

# NOTES

**Sex**

1    Frank McKinney, Scott R Derrickson and Pierre Mineau. *Behaviour* Vol. 86, No. 3/4 (1983), pp. 250–294 http://www.jstor.org/discover/1023 07/4534287?uid=3738032&uid=2&uid=4&sid=21103383096483

2    Brennan PL, Prum RO, McCracken KG, Sorenson MD, Wilson RE, *et al.* (2007). Coevolution of Male and Female Genital Morphology in Waterfowl. *PLoS ONE* 2(5): e418. http://www.plosone.org/article/ info:doi/10.1371/journal.pone.0000418

3    In Moalem, Sharon. *How Sex Works*. Harper Perennial. 2008. Page 70.

4    *Biol. Bull.* June 2007 vol. 212 no. 3 177–179 http://www.biolbull.org/ content/212/3/177.full

5    Phone interview, April 2013.

6    Ryan, Christopher and Jetha, Cacilda. 2010. *Sex at Dawn: How We Mate, Why We Stray, and What It Means for Modern Relationships.* Harper Collins.

7    Miller, Geoffrey F. 'How mate choice shaped human nature: A review of sexual selection and human evolution.' *Handbook of evolutionary psychology: Ideas, issues, and applications* (1998): 87–129.

8    *BMJ* 1999;319:1596 http://www.bmj.com/content/319/7225/1596

9    Riley AJ, Lees W, Riley EJ. An ultrasound study of human coitus. In: Bezemer W, Cohen-Kettenis P, Slob K, Van Son-Schoones N, eds. *Sex matters*. Amsterdam: Elsevier, 1992:29–36.

10  *BMJ* 1999; 319 doi: http://dx.doi.org/10.1136/bmj.319.7225.1596 (Published 18 December 1999). http://www.bmj.com/content/319/7225/1596

11  *J. Sex Marital Ther.* 2003;29 Suppl 1:71–6. http://www.ncbi.nlm.nih.gov/pubmed/12735090

12  http://www.ncbi.nlm.nih.gov/pmc/articles/PMC3845645/

13  *Current Biology*, Vol. 15, 543–548, March 29, 2005, http://www.mind.duke.edu/files/sites/platt/pub/2341148909.pdf

14  *South Med J.* 2000 Jan;93(1):29–32. Supernumerary breast tissue: historical perspectives and clinical features. http://www.ncbi.nlm.nih.gov/pubmed/10653061

15  Pseudomamma on the foot: An unusual presentation of supernumerary breast tissue. Délio Marques Conde MD, PhD1, Eiji Kashimoto MD1, Renato Zocchio Torresan MD, PhD1, Marcelo Alvarenga MD, PhD2. Dermatology Online Journal: 12 (4): 7 http://escholarship.org/uc/item/39n411b8

16  *Ann Dermatol Venereol.* 2001 Feb;128(2):144–6. [Complete supernumerary breast on the thigh in a male patient] http://www.ncbi.nlm.nih.gov/pubmed/11275593

17  *J Sex Med.* 2011 Oct;8(10):2822–30. doi: 10.1111/j.1743–6109.2011.02388.x. Epub 2011 Jul 28. http://www.ncbi.nlm.nih.gov/pubmed/21797981

18  *The Journal of Neuroscience*, 22 June 2005, 25(25): 5984–5987; doi: 10.1523/JNEUROSCI.0712–05.2005 http://www.jneurosci.org/content/25/25/5984.full

19  *J Sex Med.* 2011 Oct;8(10):2822–30. doi: 10.1111/j.1743–6109.2011.02388.x. Epub 2011 Jul 28. http://www.ncbi.nlm.nih.gov/pubmed/21797981

20  *Annu Rev Sex Res.* 2005;16:62–86. http://www.ncbi.nlm.nih.gov/pubmed/16913288

21  *European Journal of Neuroscience* Volume 24, Issue 11, pages 3305–3316, December 2006 http://onlinelibrary.wiley.com/doi/10.1111/j.1460–9568.2006.05206.x/full

22  http://www.lesleyahall.net/clitoris.htm

23  Hite, S. (1976). *The Hite report: A nationwide study on female sexuality.* New York: Macmillan.

24 Helen Rodnite Lemay, *Women's Secrets: A Translation of Pseudo-Albertus Magnus's 'De Secretis Mulierum' with Commentaries* (Saratoga Springs: State University of New York Press, 1992), 6, 130–135. Quoted in Maines, Rachel. 1999. *The Technology of Orgasm*. John Hopkins University Press.

25 Figure quoted from Roach, Mary. 2009. *Bonk: The Curious Coupling Of Sex and Science*. Canongate Books Ltd.

26 Interview: October 2013.

27 http://www.thelancet.com/themed/natsal

28 http://www.kinseyinstitute.org/resources/ak-data.html#orgasm

29 Figure quoted from Roach, Mary. 2009. *Bonk: The Curious Coupling Of Sex and Science*. Canongate Books Ltd.

30 Rathmann, WG (1959). 'Female circumcision, indications and a new technique'. *GP* 20: 115–20.

31 Figure quoted from Roach, Mary. 2009. *Bonk: The Curious Coupling Of Sex and Science*. Canongate Books Ltd.

32 Hite, S. (1976). *The Hite report: A nationwide study on female sexuality*. New York: Macmillan.

33 Ernest Gräfenberg (1950). 'The role of urethra in female orgasm'. *International Journal of Sexology* 3 (3): 145–148.

34 In Ladas, Alice Khan; Whipple, Beverly; and Perry, John D. 2005. *The G Spot and other discoveries about human sexuality*. Owl Books. Page 171.

35 Phone Interview, November 2013.

36 Amichai Kilchevsky MD, Yoram Vardi MD, Lior Lowenstein MD, Ilan Gruenwald MD. *The Journal of Sexual Medicine*. Volume 9, Issue 3, pages 719–726, March 2012 http://onlinelibrary.wiley.com/doi/10.1111/j.1743-6109.2011.02623.x/abstract

37 Adam Ostrzenski MD, PhD, Dr Hab. *The Journal of Sexual Medicine*. Volume 9, Issue 5, pages 1355–1359, May 2012. http://onlinelibrary.wiley.com/doi/10.1111/j.1743-6109.2012.02668.x/abstract

38 O'Connell HE, Hutson JM, Plenter RJ, Anderson CR. Anatomical relationship between urethra and clitoris. *J Urol*. 1998. June; 159(6):1892–1897.

39    Jannini, EA et al. *The Journal of Sexual Medicine.* Volume 5, Issue 3, pages 610–618, March 2008. http://onlinelibrary.wiley.com/doi/10.1111/j.1743–6109.2007.00739.x/abstract

40    http://www.newscientist.com/article/mg20026872.500-ecstasy-over-g-spot-therapy.html#.UeGEw-B9ndk

41    *J Sex Med.* 2008 Sep;5(9):2119–24. doi: 10.1111/j.1743–6109.2008.00942.x. Epub 2008 Jul 15.

42    Frank Addiego, Edwin G. Belzer Jr., Jill Comolli, William Moger, John D. Perry & Beverly Whipple. *Journal of Sex Research.* Volume 17, Issue 1, 1981.

43    *Annu Rev Sex Res.* 2005;16:62–86. http://www.ncbi.nlm.nih.gov/pubmed/16913288

44    Interview: May 2013.

45    Can shoe size predict penile length? J. Shah, N. Christopher. *BJU International* 13 SEP 2002 http://onlinelibrary.wiley.com/doi/10.1046/j.1464–410X.2002.02974.x/abstract

46    *International Journal of Impotence Research* (2002) 14, 283–286. doi:10.1038/sj.ijir.3900887 http://www.nature.com/ijir/journal/v14/n4/full/3900887a.html

47    *Psychology of Men & Masculinity,* Vol 7(3), Jul 2006, 129–143. http://psycnet.apa.org/?&fa=main.doiLanding&doi=10.1037/1524–9220.7.3.129

48    *European Urology* 42 (2002) 426±431. http://kozmetikcerrahi.com/indir/penissize.pdf

49    'Injection, Ligation and Transplantation: The Search for the Glandular Fountain of Youth.' Nicole L. Miller and Brant R. Fulmer. *Journal of Urology.* DOI:10.1016/j.juro.2007.01.135 http://www.urologichistory.museum/content/exhibits/historyforum/fountainofyouth.pdf

50    'Sex Gland Implantation.' G Frank Lydston. *JAMA.* 1916;LXVI(20):1540–1543. doi:10.1001/jama.1916.02580460016006. http://jama.jamanetwork.com/article.aspx?articleid=438275

51    Stanley LL: An analysis of one thousand testicular substance implants. *Endocrinology* 1922; 6: 787.

52    Figure quoted from Roach, Mary. 2009. *Bonk: The Curious Coupling Of Sex and Science.* Canongate Books Ltd.

53   *Xenotransplantation.* 2012 Nov-Dec;19(6):337–41. doi: 10.1111/xen.12004.
     Epub 2012 Oct 25 Voronoff to virion: 1920s testis transplantation and
     AIDS. http://www.ncbi.nlm.nih.gov/pubmed/23094667

54   Hamilton, David. 1986. *The Monkey Gland Affair.* Chatto & Windus.

55   *Eur J Endocrinol.* 2005 Sep;153(3):419–27. http://www.ncbi.nlm.nih.gov/
     pubmed/16131605

56   http://www.ncbi.nlm.nih.gov/pubmed/16728555

57   Volume 338, Issue 8779, 30 November 1991, Pages 1367 http://www.
     sciencedirect.com/science/article/pii/014067369192244V

58   Sourced from Roach, Mary. 2009. *Bonk: The Curious Coupling Of Sex
     and Science.* Canongate Books Ltd.

59   Laumann EO, Paik A, Rosen RC. Sexual Dysfunction in the United
     States: Prevalence and Predictors. *JAMA.* 1999;281(6):537–544.
     doi:10.1001/jama.281.6.537. http://jama.jamanetwork.com/article.
     aspx?articleid=188762

60   Magniati, Brooke. 2012. *The Sex Myth: Why Everything We're Told is
     Wrong.* W&N.

61   http://bit.ly/1hfLMoJ Sexual Behavior in the Human Male (1948).

62   Gill, KS. 1963. A mutation causing abnormal courtship and mating
     behavior in males of Drosophila melanogaster. *Am. Zool.* 3:507.

63   *Arch Gen Psychiatry.* 1991 Dec;48(12):1089–96. A genetic study of male
     sexual orientation. Bailey JM, Pillard RC. http://www.ncbi.nlm.nih.
     gov/pubmed/1845227

64   Hamer et al., 261 (5119): 321–327 http://www.sciencemag.org/
     content/261/5119/321

65   Camperio Ciani A, Pellizzari E (2012) Fecundity of Paternal
     and Maternal Non-Parental Female Relatives of Homosexual
     and Heterosexual Men. *PLOS ONE* 7(12): e51088. doi:10.1371/
     journal.pone.0051088 http://www.plosone.org/article/
     info%3Adoi%2F10.1371%2Fjournal.pone.0051088

66   *PNAS* Vol. 89, pp. 7199–7202, August 1992 Neurobiology. http://www.
     pnas.org/content/89/15/7199.full.pdf

67   Swaab DF , Hofman MA (1990) An enlarged suprachiasmatic nucleus
     in homosexual men. *Brain Res* 537:141–148

68  Wilson, G. D. & Rahman, Q. (2005) *Born Gay: The Psychobiology of Sex Orientation*. London: Peter Owen.

69  Interview: June 2013.

70  *Hormones and Behavior* 40, 105–114 (2001) http://classes.biology.ucsd.edu/bisp194–1.FA09/Blanchard_2001.pdf

71  Email Interview: November 2013.

72  *Biol Lett*. 2008 August 23; 4(4): 323–325. http://www.ncbi.nlm.nih.gov/pmc/articles/PMC2610150/

73  Phone Interview: May 2013.

74  Lesley Hall http://www.lesleyahall.net/ocbcond.htm

75  Interview: June 2011.

76  http://www.pnas.org/content/101/8/2584.1ong

77  http://ghr.nlm.nih.gov/geneFamily/hla

78  Wedekind, C., Seebeck, T., Bettens, F., Paepke, A.J., 1995. MHC-dependent mate preference in humans. *Proc. Roy. Soc. Lond.* B 260, 245–249.

79  http://rspb.royalsocietypublishing.org/content/266/1422/869.short

80  Roberts, S.C., Little, A.C., Gosling, L.M., Jones, B.C., Perrett, D.I., Carter, V., Petrie, M., 2005b. MHC-assortative facial preferences in humans. *Biol. Lett.* 1, 400–403. http://rsbl.royalsocietypublishing.org/content/1/4/400.full

81  Garver-Apgar, C.E., Gangestad, S.W., Thornhill, R., Miller, R.D., Olp, J.J., 2006. Major histocompatibility complex alleles, sexual responsivity, and unfaithfulness in romantic couples. *Psychol. Sci.* 17, 830–835.

82  *Psychoneuroendocrinology* (2009) 34, 497–512

83  Phone Interview: November 2013.

84  *Chem Senses*. 2006 Oct;31(8):747–52. Epub 2006 Aug 4. http://www.ncbi.nlm.nih.gov/pubmed/16891352

85  Interview: November 2013.

86  Frans B. M. de Waal, Our Inner Ape: The Best and Worst of Human Nature. 2006. Granta.

87  http://www.nature.com/news/gene-switches-make-prairie-voles-fall-in-love-1.13112

88  *Isr J Psychiatry Relat Sci*. 2013;50(1):33–7. http://www.ncbi.nlm.nih.gov/pubmed/24029109

89    http://www.nature.com/neuro/journal/v16/n7/full/nn.3420.html
      *Nature Neuroscience* 16, 919–924 (2013) doi:10.1038/nn.3420

90    Interview: June 2013. Note interview conducted for *Nature* http://www.
      nature.com/news/gene-switches-make-prairie-voles-fall-in-love-1.13112

91    Interview: October 2013.

92    G.G. Gallup Jr. et al. / *Evolution and Human* Behavior 24 (2003)
      277–289.

93    *Nature* 463, 801–803 (11 February 2010) | doi:10.1038/nature08736

94    http://www.jstor.org/stable/10.1086/522847

95    Lindholmer, C. (1973). Survival of human sperm in different fractions
      of split ejaculates. *Fertility and Sterility*, 24, 521–526.

96    *Biol Lett.* 2005 September 22; 1(3): 253–255. http://www.ncbi.nlm.nih.
      gov/pmc/articles/PMC1617155/

97    Ryan, Christopher and Jetha, Cacilda. 2010. *Sex at Dawn: How We
      Mate, Why We Stray, and What It Means for Modern Relationships.*
      Harper Collins.

98    http://www.queensu.ca/psychology/People/Faculty/Meredith-Chivers.
      html

**Drugs**

1     *Br J Pharmacol.* Jun 1974; 51(2): 269–278. PMCID: PMC1776744 The
      nature of the binding between LSD and a 5-HT receptor: A possible
      explanation for hallucinogenic activity M.J. Berridge and W.T. Prince

2     *J Neurosci.* 2013 Jun 19;33(25):10544–51. doi: 10.1523/
      JNEUROSCI.3007–12.2013.

3     The nicotinic acetylcholine receptor: the founding father of the
      pentameric ligand-gated ion channel superfamily Jean-Pierre
      Changeux *J. Biol. Chem.* Published online October 4, 2012 http://www.
      jbc.org/content/early/2012/10/04/jbc.R112.407668.full.pdf

4     Miller, Richard. 2014. *Drugged.* Oxford University Press.

5     Pert CB, Snyder SH. Opiate receptor: demonstration in nervous tissue.
      *Science.* 1973. 179: 1011–1014.

6     *Nature* 258, 577 – 579 (18 December 1975); doi:10.1038/258577a0 http://
      www.nature.com/nature/journal/v258/n5536/abs/258577a0.html

Identification of two related pentapeptides from the brain with potent opiate agonist activity

7    Mechoulam R, Hanus L. A historical overview of chemical research on cannabinoids. *Chem Phys Lipids.* 2000, 108 (1–2): 1–13. In Miller, Richard. 2014. *Drugged.* Oxford University Press.

8    See for example http://www.ncbi.nlm.nih.gov/pubmed/1148827

9    Wright et al. *Science.* 8 March 2013: Vol. 339 no. 6124 pp. 1202–1204 http://www.sciencemag.org/content/339/6124/1202

10   *Cellular and Molecular Life Sciences.* April 2004, Volume 61, Issue 7–8, pp 857–872. Caffeine as a psychomotor stimulant: mechanism of action. G. Fisone, A. Borgkvist, A. Usiello http://link.springer.com/article/10.1007/s00018-003-3269-3

11   See, for example, http://www.sciencedirect.com/science/article/pii/003193848890039X

12   'The company concedes to using a decocainized flavor essence in the coca leaves – one of the few Coke ingredients the company will publicly acknowledge. When asked why the company uses such a troublesome product as coca leaves, its representative said that "each ingredient adds to the flavor profile."' http://www.uic.edu/classes/osci/osci590/9_3%20The%20Legal%20Importation%20of%20Coca%20Leaf.htm

13   Transcript: http://caselaw.lp.findlaw.com/scripts/getcase.pl?court=us&vol=241&invol=265

14   Saem de Burnaga Sanchez J (1929) Sur un homologue de l'ephedrine. *Bull Soc Chim.* Fr 45:284–286

15   In Siegel, *Intoxication*, 1990, p 12.

16   In Siegel, *Intoxication*, 1990, p 24.

17   In Siegel, *Intoxication*, 1990, p 42.

18   In Siegel, *Intoxication*, 1990, p 42.

19   In Siegel, *Intoxication*, 1990.

20   In Siegel, *Intoxication*, 1990, p 51.

21   Ott, J. (1976). *Hallucinogenic Plants of North America.* Berkeley, CA: Wingbow Press. ISBN 0–914728–15–6.

22   http://www.bbc.co.uk/news/11484057

23 'A smoking monkey, who had probably mimicked the behavior of a trainer, was first displayed at a fair in The Hague in 1635.' In Siegel, *Intoxication*, 1990.

24 http://gallica.bnf.fr/ark:/12148/bpt6k64728259

25 *Psychopharmacologia* 3. 4. 1964, Volume 5, Issue 5, pp 390–392 Effect of pretreatment with tryptamine, tryptophan and DOPA on LSD reaction in tropical fish. Richard D. Chessick, Jean Kronholm, Mortimer Beck, George Maier

26 Lysergic acid diethylamide LSD-25: XXXIV. Comparison with effect of psilocybin on the Siamese fighting fish. Harold A. Abramson. *Journal of Psychology: Interdisciplinary and Applied* (1963)

27 In Siegel, *Intoxication*, 1990, p 75.

28 In Siegel, *Intoxication*, 1990, p 33.

29 *Behav Sci.* 1971 Jan;16(1):98–113. Drugs alter web-building of spiders: a review and evaluation. http://www.ncbi.nlm.nih.gov/pubmed/4937115

30 http://www.nasa.gov/multimedia/imagegallery/image_feature_629.html

31 http://epa.gov/climatechange/ghgemissions/gases/n2o.html

32 Considerations on the Medicinal Use of Factitious Airs: And on the Manner of Obtaining Them in Large Quantities. In Two Parts. Part I. by Thomas Beddoes, M.D. Part II. by James Watt, Esq. http://books.google.co.uk/books/about/Considerations_on_the_Medicinal_Use_of_F.html?id=2QcAAAAAQAAJ&redir_esc=y

33 http://archive.org/stream/humphrydavypoetpoothor/humphrydavypoetpoothor_djvu.txt

34 Miller, Richard. 2014. *Drugged*. Oxford University Press.

35 In Jay, Mike. 2000. *Emperors of Dreams*. Dedalus.

36 http://mikejay.net/articles/mushrooms-in-wonderland/ Sourced from Jay, Mike. 2010. *High Society*. Thames & Hudson.

37 Ibid.

38 See page 79 of: Hoffman, Albert. 1979. *LSD: My Problem Child*.

39 'Medicine: Mushroom Madness'. *Time*. 1958–06–16. http://content.time.com/time/magazine/article/0,9171,863497,00.html

40 *Experientia* 15. XI. 1958, Volume 14, Issue 11, pp 397–399 http://link.springer.com/article/10.1007%2FBF02160424

41  In Siegel, *Intoxication*, 1990, p 65.

42  Monatshefte für Chemie und verwandte Teile anderer Wissenschaften 1919, Volume 40, Issue 2, pp 129–154 Über die*Anhalonium*-Alkaloide. Ernst Späth.

43  http://www.youtube.com/watch?v=_wAbh3u57hA

44  See for example *Gut*, 1968, 9, 287–289. http://gut.bmj.com/content/9/3/287.full.pdf

45  See for example http://www.toxipedia.org/display/toxipedia/Ergot

46  See for example http://www.chm.bris.ac.uk/motm/lsd/lsd.htm

47  In Hoffman, Albert. 1979. *LSD: My Problem Child*, p 39.

48  *Heart.* 2003 April; 89(4): e14. Toad venom poisoning: resemblance to digoxin toxicity and therapeutic implications http://www.ncbi.nlm.nih.gov/pmc/articles/PMC1769273/

49  http://news.sciencemag.org/brain-behavior/2011/06/lsd-alleviates-suicide-headaches

50  http://clusterbusters.com/?page_id=654

51  *Neurology.* 2006 Jun 27;66(12):1920–2 Response of cluster headache to psilocybin and LSD. Sewell RA, Halpern JH, Pope HG Jr.

52  Hoffman, Albert. 1979. *LSD: My Problem Child*.

53  In Siegel, *Intoxication*, 1990, p 76.

54  In Siegel, *Intoxication*, 1990, p 76.

55  In Siegel, *Intoxication*, 1990, p 76.

56  In Siegel, *Intoxication*, 1990, p 55.

57  In Roberts, Andy. 2012. *Albion Dreaming*. Marshall Cavendish Editions.

58  http://www.amazon.co.uk/Hunter-Strange-Savage-Life-Thompson/dp/0452271290

59  'Valley of the Nerds'. *GQ* magazine, July 1991, p.96, by Walter Kirn

60  Phone interview April 11 2013.

61  Phone interview: 2012.

62  Sample papers: http://www.ralph-abraham.org/articles/

63  http://www.hippiessavedphysics.com

64  http://www.mayanmajix.com/art1699.html

65 http://blogs.nature.com/news/2011/08/weed_sequenced_no_really_ weed.html?WT.mc_id=TWT_NatureNews and http://csativa. elasticbeanstalk.com

66 http://www.amazon.com/Dancing-Naked-Mind-Field-Mullis/ dp/0679774009

67 See for example http://www.theguardian.com/uk/2006/aug/18/ topstories3.drugsandalcohol

68 See for example 3-Methoxy-4 5-methylenedioxy Amphetamine, a New Psychotomimetic Agent ALEXANDER T. SHULGIN *Nature* 201, 1120–1121 (14 March 1964) | doi:10.1038/2011120a0 http://www.nature. com/nature/journal/v201/n4924/pdf/2011120a0.pdf

69 http://www.erowid.org

70 http://www.nytimes.com/2005/01/30/magazine/30ECSTASY.html?_r=0

71 G. A. Ricaurte, J. Yuan, G. Hatzidimitriou, B. J. Cord, U. D. McCann, *Science* 297, 2260 (2002)

72 Retraction of Ricaurte et al., *Science* 297 (5590) 2260–2263. Science 12 September 2003: Vol. 301 no. 5639 p. 1479 DOI: 10.1126/science.301.5639.1479b http://www.sciencemag.org/ content/301/5639/1479.2

73 G Rogers et al. The harmful health effects of recreational ecstasy: a systematic review of observational evidence, 2009. NIHR Health Technology Assessment. http://www.ncbi.nlm.nih.gov/pubmedhealth/ PMH0014966/

74 Nutt et al. 2014. The Effects of Acutely Administered 3,4-Methylenedioxymethamphetamine on Spontaneous Brain Function in Healthy Volunteers Measured with Arterial Spin Labeling and Blood Oxygen Level–Dependent Resting State Functional Connectivity. *Biological Psychiatry*. http://dx.doi.org/10.1016/j. biopsych.2013.12.015 http://www.biologicalpsychiatryjournal.com/ article/S0006-3223(14)00005-5/abstract

75 Interview Date: April 26 2013.

76 http://drugsmeter.com

77 Hanes, KR. (2001). Antidepressant effects of the herb Salvia divinorum. *Journal of Clinical Psychopharmacology*, 21, 634–635.

78 Goldacre, Ben. 2012. *Bad Pharma*. Fourth Estate.

79    Interview: April 2013.

80    Interview in person, November 2012.

81    http://ec.europa.eu/eahc/documents/health/leaflet/hiv_aids.pdf

82    http://www.flickr.com/photos/zoecormier/sets/72157624000034891/

83    Interview in person: October 2010.

84    Siegel, Ronald. 1989. *Intoxication*. Pocket Books. P184

85    Siegel, Ronald. 1989. *Intoxication*. Pocket Books.

86    *Pharmacol Biochem Behav.* 1981 Oct;15(4):571–6. http://www.ncbi.nlm.
      nih.gov/pubmed/7291261?dopt=Abstract

87    Phone Interview: April 2013.

88    Phone Interview: March 2013. gkoob@scripps.edu

89    Krebs and Johansen. Lysergic acid diethylamide (LSD) for alcoholism:
      meta-analysis of randomised controlled trials. *J Psychopharmacol*
      March 8, 2012 http://jop.sagepub.com/content/early/2012/03/08/02698
      81112439253.abstract

90    http://www.beckleyfoundation.org/2010/09/
      psilocybin-facilitated-addiction/

91    http://www.nature.com/news/2010/100416/full/news.2010.188.html

92    *J Psychopharmacol* November 20, 2012 0269881112456611 http://jop.
      sagepub.com/content/early/2012/08/29/0269881112456611.full

93    Phone interview, March 2013. george@newmexico.com

94    Greer G, 1985. Using MDMA in psychotherapy. *Advances* 2: 57–59.

95    Sewell et al. Response of cluster headache to psilocybin and LSD http://
      www.neurology.org/content/66/12/1920.abstract

96    *J Clin Psychiatry* 67:11, November 2006.

97    Phone interview: April 2013.

98    *Biol Psychiatry.* 2000 Feb 15;47(4):351–4. http://www.ncbi.nlm.nih.gov/
      pubmed/10686270

99    Zarate, et al. *Arch Gen Psychiatry.* 2006;63(8):856–864. doi:10.1001/
      archpsyc.63.8.856. aminhttp://archpsyc.jamanetwork.com/article.
      aspx?articleid=668195

100   *Neuroscience.* 2003;117(3):697–706. Effects of ketamine and
      N-methyl-D-aspartate on glutamate and dopamine release in the rat
      prefrontal cortex: modulation by a group II selective metabotropic

glutamate receptor agonist LY379268. http://www.ncbi.nlm.nih.gov/pubmed/12617973

101  *Philos Trans R Soc Lond B Biol Sci.* 2012 Sep 5;367(1601):2475–84. A neurotrophic hypothesis of depression: role of synaptogenesis in the actions of NMDA receptor antagonists. Duman RS, *Li N.* http://www.ncbi.nlm.nih.gov/pubmed/22826346

102  *Am J Psychiatry* 2013;170:1134–1142. doi:10.1176/appi.ajp.2013.13030392

103  Phone interview: May 2013. johansenpo@gmail.com

104  Carhart-Harris, R. et al. Neural correlates of the psychedelic state as determined by fMRI studies with psilocybin. *PNAS* January 23, 2012. http://www.pnas.org/content/early/2012/01/17/1119598109.abstract

105  See for example: *Neuropsychopharmacology* (1999) **21**, 2S–8S. The Discovery of Serotonin and its Role in Neuroscience. Patricia Mack Whitaker-Azmitia Ph.D

106  *Arch Gen Psychiatry.* 2011 Jan;68(1):71–8. doi: 10.1001/archgenpsychiatry.2010.116. Epub 2010 Sep 6. http://www.ncbi.nlm.nih.gov/pubmed/20819978

107  Phone interview, March 2013. cgrob@labiomed.org

108  Griffiths et al. *J Psychopharmacol.* July 1, 2008. http://jop.sagepub.com/content/early/2008/07/01/0269881108094300.short

109  *Science.* 2009 February 13; 323(5916): 934–937. http://www.ncbi.nlm.nih.gov/pmc/articles/PMC2947205/

110  Strassman, Rick. 2001. *The Spirit Molecule.* Park Street Press.

111  Ibid.

112  DMT: The Spirit Molecule. 2010. http://www.imdb.com/title/tt1340425/

**Rock 'n' Roll**

1   Byrne, David. 2012. *How Music Works.* Canongate Books.

2   http://www.ucl.ac.uk/slms/engagement/slms-pe/case/ketamine

3   Interview Date: April 25 2013

4   http://magneticmusic.ws Magic Music From The Telharmonium. Magnetic Music Publishing Co., 1998.

5   Ibid.

6   Ibid.

7    Turk, Matija and Dimkaroski, Ljuben. 2011. 'Neandertalska piščal iz Divjih bab I: stara in nova spoznanja', 'Neanderthal Flute from Divje babe I: Old and New Findings' (English & Slovenian). *Opera Instituti Archaeologici Sloveniae : Založba ZRC SAZU, Ljubljana* 21:251–265.

8    Brodar, Mitja (26 September 2008). "Piščalka' iz Divjih bab ni neandertalska'

9    Ancient DNA Reveals Neandertals With Red Hair, Fair Complexions. *Science* 26 October 2007: Vol. 318 no. 5850 pp. 546–547 DOI: 10.1126/science.318.5850.546

10   Mithen, Steven. *The Singing Neanderthals: The Origins of Music, Language, Mind and Body.* London, Weidenfeld and Nicolson: 2005.

11   *Science.* 2010 May 7;328(5979):710–22. doi: 10.1126/science.1188021. http://www.ncbi.nlm.nih.gov/pubmed/20448178

12   *Science.* 2007 Nov 30;318(5855):1453–5. Epub 2007 Oct 25. A melanocortin 1 receptor allele suggests varying pigmentation among Neanderthals. http://www.ncbi.nlm.nih.gov/pubmed/17962522

13   Turk, Matija and Dimkaroski, Ljuben. 2011. 'Neandertalska piščal iz Divjih bab I: stara in nova spoznanja', 'Neanderthal Flute from Divje babe I: Old and New Findings' (English & Slovenian). *Opera Instituti Archaeologici Sloveniae : Založba ZRC SAZU, Ljubljana* 21:251–265.

14   Fink, Bob (1997). *Neanderthal Flute.* Greenwich. http://www.greenwych.ca/fl-compl.htm

15   Turk, Matija; Dimkaroski, Ljuben (2011). 'Neandertalska piščal iz Divjih bab I: stara in nova spoznanja' [Neanderthal flute from Divje babe I: old and new findings] http://www.cpa.si/tidldibab.pdf

16   Omerzel-Terlep, Mira. *Etnolog 6 (1996)* http://www.etno-muzej.si/files/etnolog/pdf/0354-0316_6_omerzel_koscene.pdf

17   Morphology and development of the human vocal tract: A study using magnetic resonance imaging. *J. Acoust. Soc. Am.* 106 (3), Pt. 1, September.1999.

18   Lieberman, P. 2007. The evolution of human speech; Its Anatomical and neural bases. *Current Anthropology.* 48:39–66. http://www.cog.brown.edu/people/lieberman/pdfFiles/Lieberman%20P.%202007.%20The%20evolution%20of%20human%20speech,%20Its%20anatom.pdf

19  Reznikoff, I. ' Sound resonance in prehistoric times: A study of Paleolithic painted caves and rocks'. ' *Acoustics* 08. Paris. http://webistem.com/acoustics2008/acoustics2008/cd1/data/articles/000892.pdf

20  Dauvois, M. (1989) Son et Musique Paléolithiques, *Les Dossiers D'Archéologie* Vol. 142, p. 2–11. (Read about the idea in an English review of the field here: http://www.darwin.cam.ac.uk/dcrr/dcrr002.pdf)

21  https://www.youtube.com/watch?v=LBhk5KFwLVc

22  http://guerillascience.co.uk/archives/2648 (Written by Zoe Cormier)

23  http://voyager.jpl.nasa.gov/spacecraft/music.html

24  For a bit more detail on the history of this field, check out Filler, Aaron (2010). 'The History, Development and Impact of Computed Imaging in Neurological Diagnosis and Neurosurgery: CT, MRI, and DTI'. *Internet Journal of Neurosurgery* http://precedings.nature.com/documents/3267/version/3/files/npre20093267-3.pdf?origin=publication_detail

25  *J. Acoust. Soc. Am.* 107, 2900 (2000); http://dx.doi.org/10.1121/1.428792 http://scitation.aip.org/content/asa/journal/jasa/107/5/10.1121/1.428792

26  *PLoS One.* 2008;3(10):e3511. doi: 10.1371/journal.pone.0003511. Epub 2008 Oct 29. Remote excitation of neuronal circuits using low-intensity, low-frequency ultrasound. Tyler WJ, Tufail Y, Finsterwald M, Tauchmann ML, Olson EJ, Majestic C.

27  Quam, R. M. *et al. Proc. Natl Acad. Sci. USA* http://dx.doi.org/10.1073/pnas.1303375110 (2013).

28  Coleman, M. N. & Colbert, M. W. *J. Morphol.* 271, 511–532 (2010).

29  Heffner, R., and Heffner, H. (1983). Hearing in large mammals: The horse (*Equus caballus*) and cattle (*Bos taurus*). *Behavioral Neuroscience*, 97, 299–309. http://laboratoryofcomparativehearing.com/uploads/Cattle__Bos_taurus_.pdf

30  *The Journal of Experimental Biology* 215, 3001–3009 http://jeb.biologists.org/content/215/17/3001.full.pdf

31  http://www.nature.com/news/hearing-changes-could-be-ancient-in-the-human-line-1.12976#/b1. Note the author is Zoe Cormier, and quotations from the scientists were originally used in this piece.

32    Note: Analogy derived from Daniel Levitin. http://www.newscientist.
      com/article/mg19726441.500-music-special-the-illusion-of-music.
      html#.UtxlWZE4m8o

33    *J Exp Psychol Hum Percept Perform.* 1997 Oct;23(5):1427–38. Effect of
      frequency ratio on infants' and adults' discrimination of simultaneous
      intervals. Trainor LJ. http://www.ncbi.nlm.nih.gov/pubmed/9336960

34    Quotes obtained via phone interview, January 2013. ddeutsch@ucsd.
      edu

35    You can hear a version – and several other auditory illusions – in the
      Guerilla Science Sonic Tour of the Brain, an 18-minute audio tour of
      what the brain sounds like. http://guerillascience.co.uk/archives/3750

36    Interview conducted via phone, 2008.

37    Peretz, I., Ayotte, J., Zatorre, R., Mehler, J., Ahad, P., Penhune, V. &
      Jutras, B. (2002) Congenital Amusia: A Disorder of Fine-Grained
      Pitch Discrimination. *Neuron*, vol. 33, pp. 185–191. http://www.brams.
      umontreal.ca/plab/publications/article/44

38    *Music Perception.* Vol. 25:4, PP. 357–368, 2008. http://www.nsi.
      edu/~ani/Patel_Wong_Foxton_Lochy_Peretz_2008.pdf

39    http://www.ncbi.nlm.nih.gov/pmc/articles/PMC1950825/ Peretz, I.,
      Cummings, S., & Dube, M. P. (2007). 'The genetics of congenital
      amusia (tone deafness): A family-aggregation study.' [Article].
      *American Journal of Human Genetics*, 81(3), 582–588.

40    Särkämö T, Tervaniemi M, Soinila S, Autti T, Silvennoinen HM,
      et al. (2010) Auditory and Cognitive Deficits Associated with
      Acquired Amusia after Stroke: A Magnetoencephalography and
      Neuropsychological Follow-Up Study. *PLoS ONE* 5(12): e15157.
      doi:10.1371/journal.pone.0015157 http://www.plosone.org/article/
      info%3Adoi%2F10.1371%2Fjournal.pone.0015157

41    Pinker, S. (1997). *How the Mind Works.* New York: W. W. Norton &
      Company.

42    Ibid.

43    Sperber, D. (1996). *Explaining culture: A naturalistic approach.* Oxford:
      Blackwell.

44    Levitin, Daniel. *This Is Your Brain On Music: The Science of a Human
      Obsession.* New York: Penguin, 2006.

45   Phone interview: 2008.

46   Sometimes it is described as 50%, sometimes 80%.

47   *PNAS*, September 25, 2001, vol. 98, no. 20 http://www.zlab.mcgill.ca/
     docs/Blood_and_Zatorre_2001.pdf

48   *Nature.* 2000 Mar 2;404(6773):80–4. Temporal patterns of human
     cortical activity reflect tone sequence structure. Patel AD, Balaban E.
     http://www.ncbi.nlm.nih.gov/pubmed/10716446

49   Phone interview: 2008.

50   Anatomically distinct dopamine release during anticipation
     and experience of peak emotion to music. Valorie N Salimpoor,
     Mitchel Benovoy, Kevin Larcher, Alain Dagher & Robert J Zatorre,
     January 2011 | doi: 10.1038/nn.2726. http://www.zlab.mcgill.ca/docs/
     salimpoor_2011_nn.pdf

51   Ukkola LT, Onkamo P, Raijas P, Karma K, Järvelä I (2009) Musical
     Aptitude Is Associated with AVPR1A-Haplotypes. *PLoS ONE* 4(5):
     e5534. doi:10.1371/journal.pone.0005534 http://www.plosone.org/
     article/info%3Adoi%2F10.1371%2Fjournal.pone.0005534

52   Effects of Music Listening on Cortisol Levels and Propofol
     Consumption during Spinal Anesthesia http://www.ncbi.nlm.nih.gov/
     pmc/articles/PMC3110826/

53   Blood, A. J., and Zatorre, R. J. (2001). Intensely pleasurable responses
     to music correlate with activity in brain regions implicated with reward
     and emotion. *Proceedings of the National Academy of Sciences*,98,
     11818–11823.

54   Menon, V. & Levitin, D.J. The rewards of music listening: response and
     physiological connectivity of the mesolimbic system. *Neuroimage* 28,
     175–184 (2005).

55   Phone interview, 2008.

56   http://www.youtube.com/watch?v=KzrqYPmoTVc

57   Zentner, M. & Kagan, J. (1998). Infants' perception of consonance and
     dissonance in music. *Infant Behavior and Development*, 21, 483–492.
     http://www.brainmusic.org/MBB91%20Webpage/Evolution_Zentner.
     pdf

58   *Dev Psychol.* 2001 Jan;37(1):74–85. Absolute pitch in infant auditory
     learning: evidence for developmental reorganisation. Saffran JR,
     Griepentrog GJ. http://www.ncbi.nlm.nih.gov/pubmed/11206435

59   *Music Perception* 2004 Spring 2004, Vol. 21, No. 3, 339–356. http://www.
     auditory.org/mhonarc/2004/save/pdf00001.pdf

60   *ICMPC.* 2012:618–623. Absolute Pitch and Synesthesia: Two Sides
     of the Same Coin? Shared and Distinct Neural Substrates of Music
     Listening. Loui P, Zamm A, Schlaug G. http://www.ncbi.nlm.nih.gov/
     pubmed/23508195

61   Schellenberg, E. G. (2004). Music lessons enhance IQ. *Psychological
     Science*, 15, 511– 514.

62   http://www.jneurosci.org/content/29/10/3019.1ong Krista L.
     Hyde [1] , Jason Lerch [2] , Andrea Norton [4] , Marie Forgeard [4] , Ellen
     Winner [3] , Alan C. Evans [1] , and Gottfried Schlaug [4] The Journal
     of Neuroscience, 11 March 2009, 29(10): 3019–3025; doi: 10.1523/
     JNEUROSCI.5118–08.2009

63   *Nature* 365, 611 (14 October 1993); doi:10.1038/365611a0 http://www.
     nature.com/nature/journal/v365/n6447/abs/365611a0.html

64   http://www.ncbi.nlm.nih.gov/pubmed/16597767 *Ann N Y Acad Sci.*
     2005 Dec;1060:202–9. Music listening and cognitive abilities in 10- and
     11-year-olds: the blur effect. Schellenberg EG, Hallam S.

65   Sacks, Oliver. *Musicophilia.* 2007. Picador.

66   Brain Structures Differ between Musicians and Non-Musicians
     Christian Gaser1,2 and Gottfried Schlaug. *The Journal of Neuroscience*,
     8 October 2003, 23(27): 9240–9245; http://www.jneurosci.org/
     content/23/27/9240.full

67   Cereb. Cortex (2003) 13 (9): 943–949. doi: 10.1093/cercor/13.9.943
     http://cercor.oxfordjournals.org/content/13/9/943.1ong

68   *Neuropsychologia.* 1995 Aug;33(8):1047–55. Increased corpus callosum
     size in musicians. Schlaug G, Jäncke L, Huang Y, Staiger JF, Steinmetz
     H. http://www.ncbi.nlm.nih.gov/pubmed/8524453

69   *Br J Psychiatry.* 2011 Aug;199(2):92–3. doi: 10.1192/bjp.bp.110.087494.
     http://www.ncbi.nlm.nih.gov/pubmed/21804144

70  *Pediatr Dent.* 2002 Mar-Apr;24(2):114–8. The effect of music distraction on pain, anxiety and behavior in pediatric dental patients. http://www. ncbi.nlm.nih.gov/pubmed/11991313

71  *J Music Ther.* 2006 Winter;43(4):270–94. http://www.ncbi.nlm.nih.gov/ pubmed/17348756

72  Music therapy for autistic spectrum disorder (Review) http://www. ebp-slp.com/pdfs/music-therapy/MusicTherapyandAutism.pdf

73  Interview: December 2013.

74  *J Cogn Neurosci.* 2010 Dec;22(12):2716–27. http://www.ncbi.nlm.nih. gov/pubmed/19925203

75  Racette, A., Bard, C., and Peretz, I. (2006). Making non-fluent aphasics speak: Sing along! *Brain,*129, 2571–2584.

76  Albert, M. L., Sparks, R.W., and Helm, N. A. (1973). Melodic intonation therapy for aphasia. *Archives of Neurology,* 29, 130–131.

77  Särkämö, T., Tervaniemi, M., Laitinen, S., et al. (2008). Music listening enhances cognitive recovery and mood after middle cerebral artery stroke. *Brain*, 131, 866–876. http://www.ncbi.nlm.nih.gov/ pubmed/18287122

78  *Clin Rehabil* July 2003 vol. 17 no. 7 713–722 http://cre.sagepub.com/ content/17/7/713.short

79  http://www.theguardian.com/music/2009/apr/12/ tourettes-pianist-music-performance

80  *Pediatrics* Vol. 131 No. 5 May 1, 2013 pp. 902 -918 (doi: 10.1542/ peds.2012-1367) http://www.pediatricsdigest.mobi/content/131/5/902. full

81  Interview: 2008.

82  Brains swinging in concert: cortical phase synchronisation while playing guitar. Ulman Lindenberger1*, Shu-Chen Li1, Walter Gruber2 and Viktor Mülleri. *BMC Neuroscience* 2009, 10:22 doi:10.1186/1471– 2202-10-22 http://www.biomedcentral.com/1471-2202/10/22

83  Interview: 2008.

84  *Evolutionary Psychology* 10(4): 688–702 http://www.epjournal.net/ articles/performance-of-music-elevates-pain-threshold-and-positive- affect-implications-for-the-evolutionary-function-of-music/

85    http://www.ncbi.nlm.nih.gov/pubmed/12814197 *Integr Physiol Behav Sci.* 2003 Jan-Mar;38(1):65–74. Does singing promote well-being?: An empirical study of professional and amateur singers during a singing lesson. Grape C, Sandgren M, Hansson LO, Ericson M, Theorell T.

86    Phone Interview: 2013.

87    Lipo, C. P., Hunt, T. L. & Haoa, S. R. *J. Archaeol. Sci.* http://dx.doi.org/10.1016/j.jas.2012.09.029 (2012).

88    Mithen, Steven. 2006. *The Singing Neanderthals.* Phoenix.

89    Luck, G., Saarikallio, S., Thompson, M. R., Burger, B., & Toiviainen, P. (2010). Effects of personality and genre on music-induced movement. In S. M. Demorest, S. J. Morrison, & P. S. Campbell (Eds.), *Proceedings of the 11th International Conference on Music Perception and Cognition, (ICMPC11)* (pp. 123–126). Seattle, WA: University of Washington.

90    Fink, B., Seydel, H., Manning, J.T. & Kappeler, P.M. (2006). A preliminary investigation of the associations between digit ratio and women's perception of men's dance. *Personality and Individual Differences,* 42, 381–390.

91    http://bps-research-digest.blogspot.co.uk/2006/12/mens-dancing-style-determined-in-womb.html

92    Personality and Individual Differences 50 (2011) 668–672. http://www.psychologytoday.com/files/attachments/38540/lovatt-2011.pdf

93    http://www.plosgenetics.org/article/info%3Adoi%2F10.1371%2Fjournal.pgen.0010042

94    http://www.plosone.org/article/info%3Adoi%2F10.1371%2Fjournal.pone.0005534

95    *Sports Med.* 2007;37(4–5):288–90.The evolution of marathon running : capabilities in humans. Lieberman DE, Bramble DM.

96    http://www.bbc.co.uk/programmes/p010sv12

97    Music discriminations by carp (Cyprinus carpio). *Animal Learning & Behavior,* 2001, 29(4), 336–353)

98    *Cognition* 104 (2007) 654–668. Nonhuman primates prefer slow tempos but dislike music overall.

99    Patel, A.D., J.R. Iversen, M.R. Bregman, I. Schulz & C. Schulz. 2008. Investigating the human-specificity of synchronisation to music. In Proceedings of the 10th International Conference on Music Perception

& Cognition (ICMPC10). K. Miyazaki et al., Eds.: 100–104. Causal
Productions. Adelaide, Australia.

100 Cook, P., Rouse, A., Wilson, M., & Reichmuth, C. (2013, April 1). A
California Sea Lion (Zalophus californianus) Can Keep the Beat:
Motor Entrainment to Rhythmic Auditory Stimuli in a Non Vocal
Mimic. Journal of Comparative Psychology. http://www.apa.org/pubs/
journals/features/com-128-2-1.pdf

# LIST OF ILLUSTRATIONS

While every effort has been made to contact copyright-holders of illustrations, the author and publishers would be grateful for information about any illustrations where they have been unable to trace them, and would be glad to make amendments in further editions.

# INDEX

Figures in *italics* indicate captions.

Avena, Dr Nicole 164–5, 166
AVPR1a gene 262
ayahuasca 122, 123, 177–8

**B**
BaAka people (Congo) 205, *206*
babies, and music 248–50
Bach, Johann Sebastian:
    Brandenburg Concerto No.2 208
Backstreet Boys 267
bacteria, beneficial 71–2
balance 223, 224, 245
*Banisteriopsis caapi* vine 122
barbituates/barbiturates 150, 153, 174
barley 127
Barrett, Syd 192
basilar membrane 229, 230–31
bass bins 218
bass sounds 217–21, 224, 231
bats
    common brown (*Pipistrellus
        pipistrellus*) 4*n*; ears 222;
        echolocation 129, 221; little
        brown (*Myotis lucifugus*) 4*n*;
        sonar of 13
Battersea Power Station, London xiii
Bayer 116–17
BBC Radiophonic Workshop 211
BBC television 215
Beatles 157–8, 191, 192, 213, 215, 216,
    236
'beautiful math' 232, *233*
beaver testicle moonshine 63
Beck 250
Beddoes, Thomas and Watt, James:
    *Considerations on the Medical
    Use and on the Production of
    Factitious Airs* 111

Beefheart, Captain 191, 194
beer 154
bees: memory and caffeine 100
Benzedrine 137
benzene 116
benzodiazepines 150
benzoylmethylecgonine 117, 118,
    119
Bergler, Edmund and Kroger,
    William S.: *Kinsey's Myth of
    Female Sexuality: The Medical
    Facts* 31
Berlin, Irving: 'Monkey Doodle-
    Doo' 45
Berliner, Emile 208
binomial nomenclature 4*n*
*Biological Bulletin, The* 6
*Biological Psychiatry* (journal) 173
*Biophilia* (album) 237
bipedalism 228, 263
bipolar disorder 155, 192
'birdcage man' of San Francisco
    136*n*
birds
    birdsong 264, 265, 266; pair-
        bonding 76; sacculus in 222–3;
        sensitivity to nicotine 93, 103
birth control pills 63–5, 69
birth order 59
bisexuality 88
Bjork 236–7
*Blade Runner* (film) 275
Blanchard, Professor Ray 59
'Blur effect' 252
Bob Marley & The Wailers 190
body-size dimorphism 84
Boleyn, Anne 17
Bon Jovi 267

experimental revelations
139–45, 272; experiments on
prisoners of war 124–5; first
acid trip 91, 122, 125, 131–3, *132*;
healing the mind 171–5, *172*;
influence on musicians 187–94;
legality of 93; as medicaments
167–71; in nineteenth century
115–20, *118*; nitrous oxide
110, 112–15; pharmaceutical
companies 154–7, 171; potency
120–21; renegade research
145–50; *see also* individual
drug names
Drugsmeter.com 151
druids 121
*drukQs* (album) 191
drum kits 213, 250–51, 276
drum machines
TR303 214–15; TR808 207
drumming circles 256, 257
dub 190
dubstep 190
ducklings 102
Dully, Howard *172*
Dunbar, Professor Robin 257
DVDs 210
dystimbrics 239

**E**
eardrum (tympanum) 225, 226, 228,
231, 242
ears 222, 224–31, 271
inner ear 225, 227, 229–31, *230*;
middle ear 225–8, *227*, 231; the
most basic link between sex
and rock and roll 224; outer
ear 335; sacculus 222–3, *223*,

261; shapes and forms 222;
vestibular system 223–4
Easter Island statues 258–60, *259*
ecbolics 127–8
echidna
egg-laying mammal 230*n*;
hemipenis of *80*, *81*, 82, 230*n*
echolocation 129, 221
Eclectic Medical University, Kansas
46
ecstasy 91, 145
*see also* MDMA
Edison, Thomas 207, 208
EEG (electroencephalogram)
sensors 238, 257
ejaculation 23–4, 29, 82
female 26, 37, 38–9
Elbe, Lili 32*n*
electric eels 96
electrical amplification 195
electronic music 212, 237
electronic synthesisers 197
electronica 190
elephants 103, 133, 221, 222
Ellis, Dock 137–8, 139, 158
EMI (Electrical and Musical
Industries Ltd) 215, 216
EMI Group 215
EMI Records 215
Emory University, Atlanta 79
endocannabinoids xvii, 98, 99, 180
endocrine system 224
endorphins xviii, 95, 97, 98, 180, 187,
247, 257
endrocrinology 32
English Chamber Orchestra 256
Eno, Brian 234, 235*n*
Ephron, Zac 17

'Great Starvation Experiment' 156
Green, Arda 178
Green Park, London 121
greenhouse gases 110
Greer, Dr George 169–70
Griffiths, Dr Roland 169
Grob, Professor Charles 180
'gross indecency' 53
guanine (G) 144
*Guardian* newspaper 256
Guaynerius, Athonius 26–7
Guercken, Valentina Pavlovna 122
Guerilla Science xiii-xiv, 72, 216,
    218, 262, 274–7
    Camp 274; Sonic Tour of the
    Brain 236*n*
Guevara, Che 238
Gwar *193*, 194
Gynergen 130

**H**
hagfish 97–8, *97*
Haight-Ashbury, San Francisco
    148
Halban, Josef 32
Hall, Lesley 24, 85
hallucinogens 139
    endogenous 180; ergot fungus
    127; experiments on prisoners
    of war 124–5; and fly algaric
    *102*; of the new world 116;
    and rhesus monkeys 107–9,
    *108*; and serotonin xvii, xviii;
    teenage experimenters 135;
    telescopes for the mind 176;
    and wildlife 101, 103; *see
    also* ayahuasca; LSD; magic
    mushrooms; psilocybin

Hamilton, Dr David: *The Monkey
    Gland Affair* 46–7
hands, sensory nerve endings in
    19
Hannah (a female orang-utan) 72,
    73–4, *73*
Hardison, Casey 167–8
harmonics 232
harpsichords 206
Harrison, Beatrice 264, *266*
Harvard University 117, 122, 135,
    178, 249
hash clubs 133
hashish 134
    *see also* cannabis; ganja;
    marijuana; weed
Havlicek, Dr Jan 70, 71
Hawaiian baby woodrose (*Argyreia
    nervosa*) 129
hawkmoths 102
head shops 122, 140
headaches, cluster 130, 170
*Heart of Saturday Night, The*
    (album) 194
heavy metal 202, 216, 253
hedonism 176, 224, 270, 272
    definition xvi; denial of
    hedonistic instincts xviii;
    hedonistic genes 166;
    hedonistic triad 269;
    importance of hedonistic
    impulses xix
Heffter, Arthur 124
Heisenberg, Werner 141
Hemingway, Ernest 7
Hendrix, Jimi 160, 192
hepatitis C 160
*Heroes* (album) 193